Structure of the Book

This book is organized into 14 chapters, each exploring a unique aspect of Conscienics, from foundational theories to practical applications. Beginning with scientific ideas from pioneering thinkers, the chapters progress through the empirical evidence for consciousness as a field, the neurobiological mechanisms that might underpin it, and experimental methodologies. Later sections examine real-world applications, ethical considerations, and future research directions. Additionally, two supplementary manuscripts delve into applied frameworks that extend Conscienics principles into specific fields.

Here's an overview of the 14 chapters:

Chapter 1: Foundational Theories Influencing Conscienics – Page 13
An exploration of scientific ideas from pioneering thinkers that form the foundation of Conscienics.

Chapter 2: Government Research into Consciousness – CIA Projects

and Their Relevance to Conscienics – Page 22
Examining CIA programs like MK-Ultra, Gateway Process, and Stargate Project to understand historical consciousness research.

Chapter 3: Literature Review and Theoretical Foundations – Page 28
Highlighting studies, phenomena, and analogies like the 100th Monkey Effect and cymatics supporting consciousness as a shared field.

Chapter 4: Conscienics as a Revolutionary Framework Across Science, Healthcare, and Society – Page 36
Defining the principles and structure of Conscienics as a unified theory of consciousness.

Chapter 5: Neurophysiological Mechanisms in Consciousness Resonance – Page 43
Exploring biological foundations such as mirror neurons, gamma waves, and microtubules supporting resonance within consciousness fields.

Chapter 6: Methodologies for Testing Conscienics – Page 52

Scientific approaches for validating Conscienics, including brainwave synchronization, group intention studies, and physiological coherence.

Chapter 7: Technological and Therapeutic Applications of Conscienics – Page 62
Investigating the potential for resonance-based health technologies, therapies, and consciousness-enhancing devices.

Chapter 8: Philosophical and Ethical Implications of Conscienics – Page 71
Addressing ethical considerations and the responsibilities of collective consciousness and influence.

Chapter 9: Applications of Conscienics Across Disciplines – Page 81
Exploring the impact of Conscienics on psychology, sociology, education, environmental science, and healthcare.

Chapter 10: Environmental Ethics and Influence of Consciousness on Nature – Page 91
Examining how human consciousness

may interact with and impact natural systems and ecosystems.

Chapter 11: Social Applications and Community Practices in Conscienics – Page 99
Developing practices and programs for fostering collective coherence, empathy, and community well-being.

Chapter 12: Future Directions and Research Agenda – Page 108
Proposing pathways for advancing Conscienics through interdisciplinary collaborations and empirical research.

Chapter 13: The Interplay of Structure and Experience in Consciousness – Page 118
Bridging structural and experiential models to unify scientific and personal understandings of consciousness.

Chapter 14: Introduction to the Supplementary Manuscripts – Page 127

- **A Unified Theory of Additive and Subtractive Color**
  Exploring light and color as a

metaphor for resonance and duality in Conscienics.

- **Anthrocosmia: A Multilayered Framework Integrating Human Consciousness within a Cosmic Context**
  Extending Conscienics principles to integrate personal, communal, and cosmic consciousness.

Additional Manuscripts

Two supplemental manuscripts accompany this work, offering applied frameworks to deepen readers' understanding of Conscienics:

- **A Unified Theory of Additive and Subtractive Color**: This manuscript explores light and color through additive (RGB) and subtractive (CMY) models, drawing a parallel to Conscienics' emphasis on resonance and duality. In Conscienics, consciousness is both expansive and grounded, much like how light either combines or filters to create different colors. This color theory manuscript uses the interaction of colors as a metaphor for the various frequencies of consciousness, illustrating how states can either align harmoniously or create dissonance, making abstract ideas within Conscienics visually accessible and scientifically grounded.

- **Anthrocosmia: A Multilayered Framework Integrating Human Consciousness within a Cosmic Context**: Expanding Conscienics into a cosmic framework, Anthrocosmia positions individual consciousness within universal patterns, reinforcing the idea of an interconnected system of energies. By layering personal, communal, and cosmic consciousness, this framework demonstrates how individual resonance might align with broader universal frequencies, supporting Conscienics' proposal of non-local, collective consciousness fields that connect individuals to a larger whole.

Together, these manuscripts apply Conscienics principles within specific contexts, providing readers with practical insights into how the model functions across diverse scientific and philosophical domains.

By the end of this book, you'll have a new perspective on consciousness and how interconnected we might be. Conscienics challenges us to rethink what it means to be conscious, recognize our potential influence on each other, and explore how shared resonance could shape our world. Whether you're a scientist, a spiritual seeker, or simply curious, this journey through Conscienics and its applied frameworks invites you to explore the profound web of consciousness that connects us all.

Introduction

Imagine a vast, interconnected field of energy permeating the universe, linking every person, thought, and emotion together. Conscienics is a pioneering model that postulates consciousness functions as a non-local, resonant field—a unified structure that enables minds and emotions to connect across vast distances, possibly independent of physical constraints such as space and time. Just as electromagnetic fields transmit energy over distances, this consciousness field could act as a medium that allows for interconnection on a deeply fundamental level.

In Conscienics, consciousness is not confined within individual minds; rather, it is an expansive, interconnected field that resonates through each of us. This model builds upon both ancient philosophical insights and cutting-edge scientific theories to propose a new way of understanding consciousness as a dynamic, collective field of energy and information.

Consciousness as a Field

Consider consciousness as a field akin to gravitational or electromagnetic fields—physical phenomena that exert influence over distances. However, this consciousness field is unique, posited as a medium capable of transmitting thoughts, emotions, and intentions independently of physical constraints. This non-local consciousness field connects individuals, allowing minds to interact in ways that transcend physical separation.

In Conscienics, this consciousness field is envisioned as the underlying fabric that connects not only people but potentially all matter and energy in the universe. Through this field, we are not isolated entities; we are fundamentally linked by an energetic medium that transmits our mental and emotional states.

Resonance and Vibration

In the Conscienics model, resonance plays a central role in understanding how consciousness fields might interact. Each individual mind can be thought of as resonating at a specific frequency, influenced by their mental and emotional states. This concept of resonance aligns with physical phenomena, where systems that share similar frequencies can synchronize or amplify each other's effects. Thus, in

Conscienics, thoughts, emotions, and intentions are considered forms of "mental frequencies" that can resonate within this consciousness field.

When two or more people share a similar state of mind—such as during meditation, focused intention, or emotional alignment—their consciousness fields can align, amplifying each other's mental and emotional energy. This resonance is akin to musical instruments in an orchestra tuning to the same pitch, creating a harmonic, amplified sound. Through resonance, individuals might strengthen their connections within the consciousness field, influencing each other and potentially even the physical world.

Non-Locality

Conscienics draws on quantum physics to explain how consciousness may operate independently of distance. In quantum theory, phenomena such as entanglement demonstrate non-local connections where particles remain connected and can influence each other instantaneously, regardless of distance. By applying the concept of non-locality to consciousness, Conscienics proposes that minds, too, may be able to interact across vast distances without a direct physical link.

In this view, consciousness is not bound by physical space, and the consciousness field allows for the transmission of thoughts and emotions beyond traditional physical limits. This non-local aspect of consciousness could explain phenomena where individuals feel connected to others despite being physically apart or are able to share emotions and intentions in a way that defies spatial separation.

Observer Effect and Intention

In quantum physics, the observer effect illustrates that the act of observation can influence the outcome of an experiment. Conscienics extends this principle to consciousness, proposing that focused attention or intention can amplify the resonance of the consciousness field. In this view, mental focus increases coherence within the consciousness field, enhancing its effect on others or even on physical reality.

For example, when an individual directs positive intentions toward a goal, they may be increasing the resonance of their consciousness field in a way that influences outcomes. This effect is similar to the way focused observation in quantum experiments can alter the state of particles. Through deliberate focus, individuals could potentially amplify the coherence and impact of their consciousness field, thus shaping their reality in subtle yet meaningful ways.

Collective Consciousness

Another foundational concept in Conscienics is the idea of a collective consciousness, which suggests that when enough individuals focus on the same thought, emotion, or goal, they create a "super-resonance." This amplified consciousness field could lead to synchronized changes across groups or even society. This phenomenon has parallels in studies that observe decreased stress or crime rates in cities where large groups engage in collective meditation or intention-setting practices.

In the Conscienics model, collective consciousness is seen as a powerful mechanism by which resonance within the consciousness field can grow stronger as more people participate. When groups of people meditate, pray, or focus on a shared intention, their collective focus creates a resonance bridge—an alignment within the consciousness field that can influence collective behaviors, emotions, or even social dynamics.

Why Conscienics Matters

The Conscienics model extends beyond theoretical exploration; it has practical implications for how we understand human connection, mental health, and our potential impact on the world. By recognizing consciousness as a field that connects us all, we gain insights into how our thoughts and emotions may influence not only ourselves but also others and our surroundings.

If consciousness is truly interconnected through a shared field, the implications are profound:

1. **Human Connection**: By understanding the consciousness field, we can gain a new perspective on empathy, emotional support, and connection. We may be more closely linked than we realize, creating a support network that transcends physical boundaries.
2. **Influence on Environment**: Focusing on positive intentions or peaceful thoughts could theoretically influence our environment, promoting harmony or reducing stress collectively.
3. **Collective Focus**: When people come together

with shared intentions, the effects may ripple outward, potentially creating tangible changes in communities and society at large.

Conscienics does not offer definitive answers; it invites exploration and questions. How connected are we? Can our thoughts influence others or even shape our environment? And if we share a collective consciousness, what responsibilities do we bear for the thoughts and emotions we project?

Chapter 1: Foundational Theories Influencing Conscienics

Introduction

The Conscienics model may appear innovative, but its roots extend deep into some of history's most groundbreaking scientific theories and discoveries. Over the years, scientists, philosophers, and visionaries have proposed concepts hinting at the interconnectedness of everything in the universe. Many suggested that consciousness, rather than existing solely within individual minds, might form part of a larger, shared field linking us in ways that transcend our everyday perceptions.

In this chapter, we explore the contributions of pioneering thinkers—such as Albert Einstein, Nikola Tesla, Max Planck, David Bohm, Carl Jung, and Rupert Sheldrake—whose ideas about fields, resonance, and interconnectedness underpin the Conscienics model. Drawing from these theories, Conscienics emerges as an integrative model bridging science, philosophy, and the study of consciousness.

Albert Einstein – The Unified Field and Non-Locality

Albert Einstein, one of the most influential physicists in history, spent much of his life in pursuit of a Unified Field Theory, which he believed could explain all fundamental forces, such as gravity and electromagnetism, as aspects of a single, interconnected field. This idea forms a core concept in Conscienics, extending Einstein's vision to include consciousness as part of this universal field.

Einstein also introduced the concept of non-locality, famously referring to it as "spooky action at a distance." In quantum mechanics, non-locality describes how two particles can remain connected, affecting each other instantaneously despite any physical separation. While Einstein initially expressed skepticism about this phenomenon, non-locality has since been observed and validated in quantum experiments. Conscienics builds upon this principle, proposing that consciousness, too, operates beyond the constraints of space and time, allowing minds to connect across distances in ways that reflect the non-local behavior of particles.

Relevance to Conscienics: By treating consciousness as part of a universal field and applying the concept of non-locality, Conscienics suggests that our minds may be interconnected in ways that transcend spatial limitations. This theory supports the possibility that thoughts, emotions, and intentions could travel through a consciousness field, much like particles in quantum physics, creating connections that defy physical boundaries.

Nikola Tesla – Resonance, Frequency, and the Power of Vibration

Nikola Tesla, a visionary inventor and electrical engineer, famously asserted that understanding the universe requires focusing on "energy, frequency, and vibration." He demonstrated that energy could be transmitted over great distances by matching frequencies, a process known as resonance. Tesla theorized that all matter possesses a natural frequency, and when tuned to that frequency, it achieves harmony and coherence.

Conscienics incorporates Tesla's insights on resonance, proposing that thoughts and emotions, when vibrating at similar frequencies, may "sync" across consciousness fields, strengthening mutual connections. Like musical notes played in harmony, individuals sharing aligned intentions or emotions may create resonance bridges, amplifying each other's influence within the consciousness field.

Relevance to Conscienics: Tesla's resonance principles provide a scientific foundation for Conscienics' proposal that consciousness fields can synchronize through shared frequencies. When people align in mental or emotional states, they may enhance their connection and collective influence, producing resonance bridges that bind minds together.

Max Planck – Consciousness as the Foundation of Reality

Max Planck, the founder of quantum theory, revolutionized science with his discovery that energy exists in discrete packets, or "quanta." Beyond his scientific achievements, Planck held profound views on consciousness, asserting that "consciousness is fundamental" and suggesting that the physical universe might arise from it.

Conscienics embraces Planck's belief that consciousness underlies all matter, proposing that consciousness fields operate as a foundational, interconnected structure that shapes reality. In this model, consciousness isn't merely a byproduct of physical processes but rather a primary force driving existence.

Relevance to Conscienics: Planck's perspective on consciousness as a fundamental reality aligns with Conscienics, which posits that consciousness fields form the essential framework of our interconnected existence. This foundational view supports Conscienics' hypothesis that consciousness impacts both mind and matter.

David Bohm – The Implicate Order and Hidden Connections

Physicist David Bohm advanced the idea of an "implicate order," an underlying reality in which all things are inherently connected. He proposed that the observable world, or "explicate order," unfolds from this deeper, interconnected dimension. In Bohm's view, all objects and events are part of a continuous field, and separation is an illusion.

Conscienics draws on Bohm's implicate order to propose that

consciousness fields might similarly connect individuals, transcending the limits of physical reality. By positioning consciousness as part of this underlying order, Conscienics suggests that minds resonate within a shared, interconnected consciousness field, even when they appear isolated.

Relevance to Conscienics: Bohm's implicate order provides a conceptual framework for understanding how consciousness fields could exist as a continuous, interconnected reality. In Conscienics, this interconnectedness is not only foundational but also resonates within and across individuals, supporting shared experiences and collective awareness.

Carl Jung – Collective Unconscious and Synchronicity

Psychologist Carl Jung introduced the concepts of the collective unconscious and synchronicity to describe a shared psychological field linking all minds. Jung believed that symbols, memories, and archetypes transcend individual consciousness, existing within a collective repository accessible to everyone. His work on synchronicity—meaningful coincidences that transcend cause-and-effect explanations—suggests that resonance may connect minds beyond physical interaction.

Conscienics builds upon Jung's collective unconscious, proposing that consciousness fields might carry shared information and create synchronized experiences across distances. This model suggests that when minds resonate, they access a shared consciousness, allowing for meaningful connections and influence that reflect the principles of synchronicity.

Relevance to Conscienics: Jung's theories of the collective unconscious and synchronicity align with Conscienics' idea of a shared consciousness field. By proposing that consciousness fields connect individuals at a collective level, Conscienics offers a framework for understanding how resonance may facilitate shared experiences and connections beyond physical boundaries.

Rupert Sheldrake – Morphic Resonance and Memory Fields

Biologist Rupert Sheldrake proposed the theory of morphic resonance, which suggests that memory and behavior are transmitted through invisible "morphic fields" rather than existing solely in the brain. According to Sheldrake, these fields enable individuals to access shared information across time and space, providing a means for species-wide learning.

Conscienics extends Sheldrake's morphic resonance to propose that consciousness fields store and transmit memories, emotions, and knowledge, allowing minds to resonate and connect beyond the individual level. Morphic fields provide a potential mechanism for how consciousness fields might enable non-local influence, supporting the idea that shared experiences can be transmitted across consciousness fields.

Relevance to Conscienics: Sheldrake's morphic resonance aligns with Conscienics' proposal that consciousness fields hold shared information and facilitate connections across distances. This theory provides a possible framework for understanding how thoughts and emotions might influence others through non-local interactions.

Summary of Foundational Theories

These foundational theories form the backbone of Conscienics, each offering unique insights into how consciousness might operate as a unified, interconnected field. Whether through Tesla's resonance, Einstein's unified field, Planck's fundamental consciousness, Bohm's implicate order, Jung's collective unconscious, or Sheldrake's morphic resonance, each theory contributes a crucial piece to the Conscienics model. Together, they suggest that consciousness could be more than the sum of individual minds—potentially an expansive field connecting us all, resonating to create meaningful connections and influence reality.

Chapter 2: Government Research into Consciousness – CIA Projects and Their Relevance to Conscienics

Introduction

The Conscienics model proposes that consciousness is an interconnected, resonant field capable of influencing subjective experiences and even physical reality. This concept aligns intriguingly with research from several government-funded programs, particularly those undertaken by the CIA. These programs, designed to explore altered states of consciousness and mind-matter interactions, highlight historical efforts to understand consciousness beyond individual limitations. In this chapter, we examine three notable CIA programs—MK-Ultra, the Gateway Process, and the Stargate Project—to set a foundation for understanding how Conscienics builds upon, refines, and ethically extends these early explorations into a comprehensive framework.

CIA's Consciousness Research Projects: An Overview

1. MK-Ultra

Launched in the 1950s, MK-Ultra was a series of covert experiments that investigated consciousness manipulation using drugs, hypnosis, and sensory deprivation. Aiming to influence behavior, these studies used chemical and psychological interventions to explore mind control techniques. While MK-Ultra is historically controversial due to its unethical practices, it underscored the potential for consciousness to be altered externally, suggesting that awareness could be impacted by forces beyond individual control.

The project's results raised awareness about the mind's malleability, albeit through ethically questionable means. Despite its dark legacy, MK-Ultra demonstrated that the mind could be influenced from outside itself, a notion relevant to the Conscienics model, which emphasizes the capacity of consciousness fields to influence and harmonize with each other. However, unlike MK-Ultra, Conscienics centers on ethical engagement and mutual resonance rather than control or manipulation.

2. The Gateway Process

Developed in the late 20th century, the Gateway Process explored techniques designed to expand consciousness through the synchronization of brainwave frequencies. Utilizing Hemi-Sync audio technology, which aligns brain hemispheres using sound, this program promoted states of coherence associated with increased awareness, altered states, and even out-of-body experiences. The Gateway Process laid a scientific foundation for understanding how resonance, through frequency alignment, could enhance consciousness.

This alignment of brain hemispheres through frequency manipulation parallels the Conscienics concept of resonance within consciousness fields. While the Gateway Process used sound to foster brainwave synchronization, Conscienics proposes that this resonance effect extends across frequencies, linking minds through conscious focus, shared intentions, or emotions. The Gateway Process's focus on non-invasive techniques to expand consciousness aligns with Conscienics' ethically grounded approach to enhancing awareness.

3. Stargate Project

The Stargate Project, spanning from the 1970s to the 1990s, sought to explore remote viewing—the ability to perceive distant or unseen targets purely through conscious focus. Remote viewers were trained to access information about distant locations without physical contact or cues. Results from Stargate demonstrated instances of successful remote viewing, lending empirical support to the idea of consciousness as a field that transcends physical limitations.

The Stargate findings are foundational for Conscienics, supporting the notion that consciousness operates within a shared, non-local field. In the Conscienics model, this interconnected consciousness field allows information to flow independently of spatial constraints. Stargate provides empirical evidence that non-local awareness is possible, suggesting a functional aspect of consciousness that Conscienics interprets as an intrinsic characteristic of consciousness fields.

Aligning CIA Research with Conscienics Principles

The CIA programs explored distinct facets of consciousness that resonate with key elements of Conscienics. Each project—whether by inducing altered states, exploring remote perception, or seeking mind control—examined consciousness beyond individual boundaries. In synthesizing the lessons from these programs, Conscienics gains historical grounding while also setting a more ethical, coherent path forward.

1. **Resonance and Coherence** (Insights from the Gateway Process)

The Gateway Process's use of brainwave synchronization through sound frequencies parallels Conscienics' principle of resonance. In Conscienics, resonance extends beyond audio frequencies, encompassing the mental, emotional, and potentially even physiological domains. This creates a space where coherence can develop across consciousness fields, establishing a foundation for individuals to connect through shared resonance. By broadening the resonance principle, Conscienics proposes that shared focus, intention, or emotion can create aligned states, strengthening interpersonal connections within the consciousness field.

1. **Non-Local Consciousness** (Building on the Stargate Project)

Stargate's experiments in remote viewing suggest that consciousness can transcend spatial limits. Conscienics adopts this notion, proposing that consciousness fields exist non-locally, allowing interactions and awareness to extend beyond physical confines. Where Stargate viewed remote viewing as a skill to be honed, Conscienics considers non-locality as an innate feature of consciousness, allowing minds to resonate within a shared field regardless of location. Conscienics builds on Stargate's isolated results, proposing a reproducible, field-based model of non-local consciousness, suggesting that consciousness naturally exists within a boundless field.

1. Ethical Standards in Consciousness Research (Lessons from MK-Ultra)

MK-Ultra serves as a critical reminder of the ethical risks associated with manipulating consciousness. Unlike MK-Ultra's ethically compromised approach to altering states of mind, Conscienics emphasizes the importance of autonomy, ethical exploration, and voluntary engagement. In Conscienics, resonance and coherence practices are applied to foster alignment rather than control. Ethical principles in Conscienics advocate for personal empowerment, collective growth, and mutual consent, positioning the model as a responsible framework for exploring consciousness.

Conscienics as a Modern Evolution of CIA Consciousness Research

Conscienics builds upon and advances the insights gleaned from government research, offering a modern evolution that aligns with scientific inquiry while honoring ethical boundaries. While MK-Ultra, Gateway, and Stargate pioneered new understandings of consciousness, they often lacked cohesive frameworks or ethical standards. Conscienics, however, synthesizes these insights into a comprehensive model that respects consciousness's complexity, extends its boundaries, and adheres to ethical guidelines.

- **Integrative Resonance Practices**: Conscienics incorporates the Gateway Process's findings on coherence through brainwave synchronization and extends them. By integrating a broader range of resonant practices—including light, sound, and biofield resonance—Conscienics creates a multidimensional approach to expanding consciousness ethically and scientifically.
- **Field-Based Non-Local Consciousness**: Conscienics uses Stargate's empirical findings to support a theory of non-local consciousness. It

suggests that individual minds are naturally linked within a shared field, a structure that enables awareness and influence beyond conventional space and time limitations. Conscienics formalizes this idea, proposing a structured, field-based approach to non-local awareness, allowing for reproducible and ethically aligned explorations of consciousness.

- **Commitment to Ethical Research**: MK-Ultra serves as a cautionary tale that informs Conscienics' ethical principles. Rather than seeking to control consciousness, Conscienics emphasizes individual agency, alignment, and ethical growth. By promoting voluntary engagement in consciousness-expanding practices, Conscienics emerges as a model of ethically sound, responsible research.

Conclusion

This chapter has explored how Conscienics aligns with and builds upon historical consciousness research by examining CIA programs like MK-Ultra, the Gateway Process, and the Stargate Project. These programs serve as both foundational studies and cautionary tales, illustrating the need for ethical integrity in consciousness research. By integrating insights from these projects, Conscienics not only advances the understanding of consciousness as a non-local, resonant field but also sets a new standard for ethical exploration. Through the principles of resonance, coherence, and non-locality, Conscienics provides a scientifically grounded and ethically responsible model of consciousness that respects individual autonomy while fostering collective harmony.

In the following chapter, we will delve into existing studies on consciousness fields and explore analogies that make complex ideas within Conscienics more accessible. By reviewing empirical research and natural phenomena, we will build a case for Conscienics as a model with substantial scientific backing.

Chapter 3: Literature Review and Theoretical Foundations

Introduction

The Conscienics model proposes that consciousness operates as a resonant, non-local field capable of connecting individuals and potentially influencing physical reality. While this concept may seem abstract, a growing body of research from quantum mechanics, psychology, neuroscience, and consciousness studies provides a foundation for understanding how such a field could exist. This chapter presents key studies, theoretical frameworks, and analogies that support the ideas behind Conscienics, showing that this model aligns with and builds upon existing scientific and philosophical knowledge.

To make these complex ideas accessible, the chapter introduces analogies—such as the 100th Monkey Effect and cymatics—that illustrate how resonance, non-locality, and shared consciousness may work in practice. Together, the research and analogies presented here offer a solid foundation for understanding how Conscienics bridges scientific inquiry and evolving theories of consciousness.

Quantum Mechanics and Non-Locality: Evidence for Connection Beyond Distance

Quantum mechanics is a field that has introduced groundbreaking ideas supporting non-local connections—concepts central to Conscienics. Specifically, the phenomenon of quantum entanglement demonstrates that particles can remain connected, or "entangled," across vast distances, influencing each other instantaneously, regardless of spatial separation. Experiments by physicists such as Alain Aspect in the 1980s confirmed entanglement, showing that particle interactions transcend conventional limitations of space and time.

The implications of quantum non-locality extend beyond physics into the realm of consciousness. David Bohm, a prominent physicist, suggested that consciousness itself might operate at a quantum level, facilitating non-local interactions between minds. Bohm's ideas align with Conscienics' proposition that consciousness fields are interconnected, potentially allowing thoughts, emotions, and intentions to resonate and influence others beyond physical boundaries.

Research Highlights:

1. **Alain Aspect's Experiments on Quantum Entanglement**: Aspect's work demonstrated that particles maintain instantaneous connections over distances, providing a framework for understanding non-local interactions that may also apply to consciousness.
2. **David Bohm's Theory of the Implicate Order**: Bohm proposed that consciousness might be part of a "hidden order," where seemingly separate entities are

united by deeper, non-local connections. This theory underpins the idea of a unified consciousness field, a concept central to Conscienics.

In Conscienics, if particles can exhibit non-local behavior, consciousness—structured as a field—might also operate beyond spatial constraints. This allows for the potential of thought and emotion transmission independent of physical distance, creating a shared space for interaction.

Psychology and Collective Consciousness: Insights into Group Influence and Resonance

Psychological studies also offer insights into collective consciousness and resonance, particularly in how groups influence individuals. Carl Jung's concept of the collective unconscious suggests that certain memories, symbols, and archetypes are universally shared across humanity, existing within a shared psychological field. Jung's work points to an interconnected realm where individual consciousness taps into a collective reservoir, a concept mirrored in Conscienics' emphasis on shared consciousness fields.

One phenomenon that supports this idea is the "100th Monkey Effect," a concept describing how learned behaviors appear to spread through populations without direct contact. Although controversial, this effect illustrates how collective resonance might allow knowledge or behaviors to propagate within a shared field of consciousness.

Research Highlights:

1. **Carl Jung's Collective Unconscious**: Jung's theories suggest that individual consciousness may access shared symbols and memories, indicating a collective aspect of consciousness that aligns with Conscienics.

2. **100th Monkey Effect**: Although debated, this concept proposes that behaviors learned by individuals within a group can eventually spread across the species through collective resonance, supporting the idea of a non-local consciousness field.

Through these psychological lenses, Conscienics frames consciousness fields as interwoven and capable of influencing shared behaviors, emotions, and ideas, facilitating a non-local connection among individuals.

Neuroscience and Brain Coherence: Biological Evidence for Resonance

In neuroscience, studies on brainwave coherence provide biological evidence supporting resonance within Conscienics. Brain coherence refers to the synchronization of brainwave activity between different brain regions, often occurring during activities like meditation or deep concentration. This coherence fosters a "whole-brain" state where the mind achieves heightened awareness, clarity, and unity.

Gamma waves, high-frequency brainwaves associated with deep mental focus, play a particularly interesting role. Research has shown that gamma waves can synchronize different brain regions, potentially allowing individuals to connect more deeply with others or experience a heightened state of collective consciousness.

Research Highlights:

1. **Gamma Wave Synchronization**: Studies suggest that gamma waves allow for synchronized brain activity, promoting a unified mental state and possibly supporting Conscienics' idea of resonance bridges across individuals.

2. **Brain Coherence in Meditation**: Meditation practices increase coherence in brainwave patterns, creating aligned states that could enhance connectivity in a shared consciousness field.

Conscienics posits that such brain coherence may act as a biological mechanism for resonance, allowing individuals to connect through shared mental states. This supports the theory that the brain's natural coherence capabilities contribute to non-local connections within consciousness fields.

Cymatics and Resonance: Visualizing Sound and Frequency

Cymatics, the study of visible sound patterns, offers an accessible analogy for understanding resonance within Conscienics. By placing particles on a vibrating surface, cymatics demonstrates how different frequencies create distinct patterns, showing how resonance shapes matter. This phenomenon illustrates that frequencies can affect physical forms, a concept parallel to how resonance in consciousness fields may influence thoughts, emotions, or even physical reality.

Masaru Emoto's research on water crystallization provides a similar example, suggesting that thoughts, words, and intentions influence the crystalline structure of water. While controversial, Emoto's work underscores the potential for resonance to affect physical forms, a core principle in Conscienics.

Research Highlights:

1. **Cymatics by Hans Jenny**: Jenny's work visualizes how sound frequencies create distinct patterns, suggesting that resonance shapes physical structures and might apply to

consciousness fields as well.
2. **Masaru Emoto's Water Crystals**: Emoto's studies propose that human intention can influence water crystal formation, supporting Conscienics' idea that consciousness fields may shape reality.

Through these analogies, cymatics and Emoto's work help illustrate how resonance within consciousness fields might operate similarly, affecting both mental states and material forms.

Field Theory and Biofields: The Role of Electromagnetic Fields in Human Interaction

Biofield science explores how electromagnetic fields within and around the body may interact with the environment and other people. Studies have shown that biofields—subtle energy fields emitted by living beings—can be influenced by intentional focus, supporting the idea that consciousness operates within an interactive, resonant field.

Techniques like transcranial magnetic stimulation (TMS) demonstrate how external fields influence cognition, providing evidence that human energy fields interact with consciousness. Although biofield science is still emerging, its findings align with Conscienics by suggesting that consciousness fields resonate and influence the mind and body.

Research Highlights:

1. **Biofield Research**: Studies show that biofields are influenced by intentional focus, supporting the existence of interactive consciousness fields.
2. **Electromagnetic Field Effects on Cognition**: Research on TMS and similar techniques demonstrates that external fields can alter

mental states, lending support to Conscienics' premise of an interactive consciousness field.

In Conscienics, biofields provide a potential framework for understanding how consciousness fields might be structured and interact, influencing mental and physical states through resonance.

Summary of Theoretical Foundations for Conscienics

This chapter provides a foundational review of research supporting the core principles of Conscienics. From quantum non-locality to biofield studies, each theory contributes a unique element to the model, reinforcing the idea that consciousness is interconnected, resonant, and capable of influence beyond individual minds. Conscienics builds on these scientific insights to propose a consciousness field that connects and resonates through shared thoughts, emotions, and intentions.

In the next chapter, we'll examine empirical evidence and case studies that further illustrate the Conscienics model. By reviewing specific experiments and observed phenomena, we will explore how Conscienics might manifest in real-world contexts, bridging theory with observable experience.

Chapter 4: Conscienics as a Revolutionary Framework Across Science, Healthcare, and Society

Introduction

The Conscienics model represents an expansive framework that posits consciousness as a resonant, interconnected field capable of influencing physical and psychological realms. This chapter delves into how Conscienics aligns with, completes, and expands upon existing theories in quantum consciousness, biophoton theory, holistic healthcare, and societal unity. By integrating consciousness as an influential field within the fabric of reality, Conscienics offers a revolutionary perspective that could profoundly impact science, healthcare, psychology, and social systems, potentially fostering a new understanding of unity and well-being.

1. Quantum Consciousness Theories: Bridging the Mind-Body Gap

Quantum consciousness theories, like those of physicist Roger Penrose and anesthesiologist Stuart Hameroff, propose that consciousness arises from quantum processes within brain structures, possibly existing independently of physical matter. These theories often suggest that consciousness operates beyond conventional neural functions, engaging with quantum phenomena in ways not yet fully understood.

Conscienics Contribution: Conscienics extends quantum consciousness theories by positioning consciousness as a non-local, resonant field that interacts with matter. Through resonance and coherence, Conscienics proposes that consciousness naturally influences physical phenomena, bridging the gap between the physical and mental realms. By applying a coherent model of resonance, Conscienics suggests that thoughts and intentions, when aligned, can influence outcomes across the quantum and macroscopic worldson Theory and Energy Medicine: Toward a New Healthcare Paradigm

Biophoton theory proposes that cells in living organisms emit low-level light (biophotons), potentially facilitating cellular communication. While this theory remains on the periphery of mainstream medicine, research on biophotons suggests they might play a significant role in health and healing. Additionally, energy medicine, which includes practices like Reiki and Therapeutic Touch, uses the idea of an energy field to promote healing by aligning subtle energies within the body.

Conscienics Contribution: By framing consciousness as a resonant energy field, Conscienics offers a model that could integrate biophoton theory within healthcare. In this paradigm, health is a state of biofield coherence, with consciousness influencing cellular function through resonant frequencies. Conscienics emphasizes preventative healthcare focused on aligning consciousness and biofields, encouraging non-invasive, frequency-based therapies that reduce reliance on pharmaceuticals and promote holistic well-being .

3. Theorsness and Social Unity

Many psychological and philosophical theories suggest that humanity shares a collective consciousness that transcends individual awareness. While traditional science has not widely adopted this viewpoint, it is reflected in practices like group meditation, where participants report experiencing unity and interconnectedness. Studies from the Global Consciousness Project, which measures deviations in random number generators during global events, imply that collective emotional responses might influence physical systems .

Conscienics Contribution: Conscienics otifically grounded framework for understanding collective consciousness as a shared resonant field. By validating and measuring coherence in this field, Conscienics could inspire social systems that prioritize empathy, unity, and mutual respect. Implementing resonance-based education or community-building practices may foster collective coherence, ultimately supporting social unity and well-being.

4. Holistic Healthcare and Self-Healing Models: Expanding Medical Horizons

Alternative and holistic health models emphasize the body's self-healing abilities and the role of consciousness in physical health. Although these approaches are supported by studies on mindfulness, meditation, and heart coherence, they often lack broad scientific validation within conventional medicine.

Conscienics Contribution: Conscienics expands upon holistic health concepts by providing a scientifically plausible framework in which consciousness alignment promotes physical healing. In this model, coherence within consciousness fields fosters cellular harmony, aligning biological processes with conscious intentions. Such a shift could lead to more integrative health practices, where therapeutic methods address physical, emotional, and consciousness-related dimensions, reducing dependence on pharmaceuticals and invasive procedures .

Potential for Transforming Healthcare and Soel has the potential to address societal and healthcare challenges by redefining health, psychology, and social systems in terms of resonance, coherence, and shared consciousness.

1. **Healthcare Transformation**: A Conscienics-based approach shifts healthcare from symptom management to preventative and resonance-based health practices. This model encourages biofield alignment, coherence

exercises, and frequency-based treatments as viable methods for maintaining well-being. Such an approach could reduce pharmaceutical dependence, focusing instead on natural, resonance-centered health practices.

2. **Social and Economic Impact**: Conscienics' principles of interconnected consciousness could inspire social structures that promote empathy, shared goals, and collective resilience. Incorporating collective resonance practices, educational programs focused on empathy, and community mindfulness exercises could strengthen

community bonds, reducing social division and fostering cooperative societies.
3. **Global Unity**: By encouraging individuals to view themselves as part of a larger, interconnected consciousness, Conscienics promotes unity beyond geographical and cultural boundaries. This perspective has the potential to diminish competitive, isolationist attitudes, fostering global cooperation and environmental responsibility through shared understanding .

Conclusion: Toward a Unified Paradigm

The Conscienics model does moeories; it offers a comprehensive framework for understanding consciousness, health, and unity. By shifting paradigms toward resonance, coherence, and collective consciousness, Conscienics invites a future in which science, healthcare, psychology, and social structures operate from principles of unity, empowerment, and respect for consciousness as a foundational force.

In the next chapter, we will examine empirical evidence that further supports the Conscienics model. Through case studies and experimental findings, we will explore how resonance, coherence, and non-local connections manifest in real-world scenarios, bridging theory and practice.

Chapter 5: Neurophysiological Mechanisms in Consciousness Resonance

Introduction

The Conscienics model suggests that consciousness operates as a resonant field, connecting individuals through shared thoughts, emotions, and mental states. But what underlying biological mechanisms might enable this interconnectedness? Recent discoveries in neuroscience, quantum biology, and psychophysiology indicate that the human brain and body may possess innate structures that support resonance within a shared consciousness field. In this chapter, we explore how key mechanisms, such as mirror neurons, gamma waves, microtubules, and heart-brain coherence, provide a neurophysiological basis for resonance bridges and non-local connections proposed by Conscienics.

These insights suggest that resonance, empathy, and shared experiences are not just abstract concepts but phenomena with measurable foundations within our bodies. By examining these mechanisms, we gain insight into how the brain and body might support resonance within consciousness fields, allowing for connection and influence beyond individual minds.

Microtubules and Quantum Consciousness: A Cellular Foundation for Non-Locality

Microtubules, microscopic tubular structures within neurons, serve as channels for cellular communication and help maintain cell structure. Neuroscientist Stuart Hameroff and physicist Roger Penrose have proposed that microtubules might function as quantum processors, enabling the brain to operate at a quantum level. Their theory, known as Orchestrated Objective Reduction (Orch-OR), suggests that consciousness may arise from quantum processes within microtubules, allowing neurons to communicate in ways that are not restricted by physical boundaries.

In Conscienics, microtubules could act as "antennas" that allow neurons to resonate with a consciousness field, much like how radios pick up distant signals. If these quantum interactions are truly occurring within microtubules, they could create a biological mechanism for non-locality, providing a means for individuals to connect and share information beyond their immediate surroundings.

Key Insight: Microtubules may act as quantum interfaces that allow the brain to access and transmit information through a shared consciousness field. This quantum coherence provides a foundation for the non-local connections proposed by Conscienics, where thoughts and emotions resonate across physical distances.

Gamma Waves: High-Frequency Brainwaves Supporting Collective Resonance

Gamma waves, high-frequency brainwaves, are associated with heightened mental states, such as focus, insight, and intense awareness. During deep meditation or moments of clarity, gamma wave activity increases, fostering a sense of unity and interconnectedness. Researchers believe that gamma waves enable different brain regions to synchronize, creating a cohesive experience of consciousness.

Conscienics suggests that gamma waves might play a central role in connecting individual minds to the consciousness field. When individuals experience heightened awareness, their brains produce gamma waves that could enhance resonance bridges, creating shared mental states or collective experiences. This synchronization allows individuals to resonate within the consciousness field more deeply, making it easier to connect with others.

Key Insight: Gamma waves may facilitate resonance bridges within the consciousness field, enabling shared experiences and heightened states of awareness that support collective consciousness as proposed by Conscienics.

Mirror Neurons: Biological Mechanisms for Empathy and Resonance

Mirror neurons, specialized brain cells, activate when we observe others' actions or emotions, allowing us to "mirror" their experiences. These neurons are believed to play a central role in empathy, enabling us to understand and share others' emotions as if they were our own. This mirroring effect allows individuals to feel connected without words or direct interaction.

In Conscienics, mirror neurons provide a neurobiological foundation for resonance bridges. When individuals resonate with each other emotionally, mirror neurons create empathy-based bridges within the consciousness field, facilitating a non-verbal alignment between minds. This resonance allows us to experience shared consciousness, enabling mutual influence and deeper connection within the consciousness field.

Key Insight: Mirror neurons support natural empathy and resonance, providing a mechanism for creating consciousness bridges based on shared mental and emotional states, as described by the Conscienics model.

Brainwave Synchronization: Tuning into Shared Consciousness

Brainwave synchronization, or neural entrainment, occurs when individuals' brainwaves align during shared activities or intense emotional connections. This synchronization is often observed in group meditation, shared storytelling, or focused collaborative tasks. Neural entrainment allows individuals to experience "in sync" mental states, fostering connections that transcend spoken words.

According to Conscienics, brainwave synchronization is a way for individuals to "tune in" to each other's consciousness fields, creating resonance bridges. When people share a mental focus or emotional state, their brainwaves align, enabling them to resonate and communicate at a deeper level. This shared mental state forms the basis for resonance-based interactions, allowing individuals to feel deeply connected even without physical or verbal cues.

Key Insight: Brainwave synchronization allows individuals to resonate naturally, forming resonance bridges that create a shared consciousness state within the Conscienics framework.

Heart-Brain Coherence: The Power of Emotional Resonance

Beyond brainwaves, the concept of heart-brain coherence provides additional support for resonance within consciousness fields. Research by the HeartMath Institute has demonstrated that positive emotions, such as love, gratitude, and appreciation, lead to stable, coherent heart rhythms. This coherence can also synchronize with brainwaves, creating harmony between the heart and mind. Importantly, when individuals in group settings focus on positive emotions, their physiological coherence often aligns, suggesting that shared emotional states create resonance within the consciousness field.

In Conscienics, heart-brain coherence could amplify resonance bridges, strengthening the consciousness field and enhancing non-local connections. This phenomenon indicates that emotional states not only influence our personal experience but may also extend outward, allowing for collective resonance that unites individuals.

Key Insight: Heart-brain coherence enables emotional resonance within the consciousness field, amplifying collective coherence and enhancing the strength of resonance bridges in shared emotional states.

Quantum Coherence in the Brain: The Role of Whole-Brain Resonance

Quantum coherence, a state in which particles maintain harmony and function as a single unit, might play a crucial role in how the brain processes information. Some researchers suggest that the brain may use quantum coherence to achieve a unified consciousness experience that extends beyond individual neurons.

Within the Conscienics model, quantum coherence could enable different brain regions to resonate in a "whole-brain" state, supporting non-local connections. If the brain achieves coherence at the quantum level, this could provide a biological mechanism for shared mental states and resonance within the consciousness field.

Key Insight: Quantum coherence in the brain may provide a mechanism for whole-brain resonance, supporting the Conscienics model of non-local connections and resonance bridges within consciousness fields.

Summary: Neurophysiological Foundations for Conscienics

The structures and processes explored in this chapter provide possible biological mechanisms that support the principles of Conscienics. From microtubules and brainwave synchronization to heart-brain coherence and gamma waves, each mechanism offers insights into how our brains and bodies might participate in a broader consciousness field. These findings suggest that resonance, non-locality, and collective consciousness could be grounded in the biology of our brains and bodies.

In the next chapter, we will transition from theory to practice, examining methodologies and empirical foundations for testing Conscienics. Through experimental approaches and scientific tools, we'll explore how the Conscienics model could be validated and expanded, opening pathways for future research.

Chapter 6: Methodologies for Testing Conscienics

Introduction

For Conscienics to move from theory to validated science, empirical testing and rigorous methodological foundations are essential. This chapter explores the experimental approaches and tools necessary to examine Conscienics, offering methods to test the existence and impact of consciousness fields, resonance bridges, and non-local interactions. By integrating neuroscience, psychology, and quantum biology, these proposed studies aim to provide empirical support for the Conscienics model, paving the way for future research and practical applications.

1. Brainwave Synchronization Studies: Investigating Resonance Bridges

One of the primary ways to test resonance bridges in Conscienics is by measuring brainwave synchronization between individuals engaging in shared activities or focused attention, such as meditation or collective visualization. This method seeks to determine if shared mental states can synchronize brainwave patterns, supporting the Conscienics idea of resonant connection.

Proposed Method:

1. Recruit participants, dividing them into pairs or small groups.
2. Instruct participants to engage in shared focus exercises like meditation or visualization.
3. Use EEG (electroencephalography) to measure brainwave patterns, assessing synchronization between individuals.
4. Analyze whether groups with shared intentions show higher levels of brainwave synchronization than control groups.

Expected Outcome: Participants with shared mental focus should demonstrate increased brainwave synchronization compared to individuals without shared focus, providing evidence for resonance bridges.

2. Heart-Brain Coherence and Group Meditation: Testing Collective Resonance

Group meditation or shared intention exercises provide an ideal setting for studying collective resonance. By monitoring physiological coherence, such as heart rate variability (HRV) and brainwave coherence, in participants engaged in group meditation, researchers can assess whether positive emotions like gratitude or compassion enhance collective resonance.

Proposed Method:

1. Conduct group meditation sessions with in-person and remote participants to explore non-local effects.
2. Use biofeedback devices and EEG to monitor HRV and brainwave

coherence in participants.
3. Compare coherence levels during group meditation versus individual meditation sessions.

Expected Outcome: Group meditation should result in higher levels of heart-brain coherence among participants than individual meditation, suggesting that collective resonance supports the consciousness field.

3. Random Event Generator Experiments: Detecting Collective Consciousness

Building on the methodology of the Global Consciousness Project, random event generators (REGs) can be used to detect deviations from randomness during significant group focus events. By analyzing these deviations, researchers may identify field effects indicative of collective consciousness.

Proposed Method:

1. Place REGs at multiple locations, both near the meditation event and globally, if feasible.
2. Monitor randomness during synchronized group activities like global meditations or mass prayers.
3. Analyze deviations in randomness, comparing baseline readings with those recorded during the event.

Expected Outcome: Significant deviations in randomness during high-focus events would support the notion that collective consciousness fields can affect physical systems.

4. Microtubule and Quantum Coherence Studies: Exploring Quantum Effects in Consciousness

To examine whether microtubules within neurons operate as quantum processors, researchers can study microtubule behavior during altered states of consciousness, such as meditation. If consciousness operates at a quantum level, measurable effects on microtubule organization may provide evidence for Conscienics' resonance-based connections.

Proposed Method:

1. Conduct meditation or focused attention exercises with participants.
2. Utilize brain imaging or other advanced methods to assess microtubule activity during altered consciousness states.
3. Analyze microtubule behavior in comparison to baseline, exploring quantum coherence patterns.

Expected Outcome: Changes in microtubule patterns during altered states would suggest quantum coherence as a foundation for consciousness, supporting non-local connections proposed by Conscienics.

5. Group Intention Studies: Testing the Observer Effect on Physical Systems

Group intention studies could explore whether collective focus influences physical systems, such as water crystallization or plant growth, thereby demonstrating the observer effect in a controlled setting.

Proposed Method:

1. Use samples such as water, plants, or biological tissues as test subjects.
2. Instruct groups to focus on specific intentions, like promoting growth or enhancing structural patterns.
3. Compare test results with control samples not subjected to focused intention.

Expected Outcome: Observable differences between test and control samples would suggest that group intention influences physical matter, reinforcing the Conscienics principle of resonance-based influence.

6. Physiological Coherence in Remote Groups: Testing Non-Local Resonance

By measuring physiological coherence, such as HRV and brainwave synchronization, in remote groups, researchers can investigate whether resonance bridges form beyond physical proximity.

Proposed Method:

1. Instruct remote groups to meditate or focus on a shared goal simultaneously.
2. Measure physiological coherence using HRV and EEG, comparing remote and in-person groups.

3. Evaluate whether coherence levels remain comparable despite distance.

Expected Outcome: Similar coherence levels between remote and in-person groups would support the theory of non-local resonance within consciousness fields.

7. Collective Resonance and Environmental Influence: Studying Group Impact on Environmental Systems

Some research suggests that human consciousness may impact environmental factors like atmospheric conditions or biological organisms. Researchers can monitor environmental variables during large group focus events to determine if collective resonance influences local conditions.

Proposed Method:

1. Organize large-scale focus events and

monitor atmospheric or biological variables, such as air quality or plant growth.
2. Record environmental changes during the event and compare with baseline data.
3. Analyze for significant deviations that correlate with the group focus.

Expected Outcome: Detectable environmental changes during focus events would indicate that collective consciousness can influence environmental systems, supporting the Conscienics model.

Summary: A Roadmap for Validating Conscienics

These experimental approaches provide a structured path for testing Conscienics in real-world settings. Whether through brainwave synchronization, physiological coherence, random event generators, or environmental impact studies, these methods offer a comprehensive framework for investigating consciousness fields and resonance bridges. Empirical validation of these effects would solidify Conscienics as a scientifically grounded model, opening the door to new insights and practical applications.

In the next chapter, we'll explore the practical applications of Conscienics, focusing on technologies and therapeutic methods that could leverage consciousness fields and resonance for health, well-being, and personal growth.

Chapter 7: Technological and Therapeutic Applications of Conscienics

Introduction

The Conscienics model suggests that consciousness operates as an interconnected, resonant field, opening up exciting possibilities for technology and therapy. By leveraging resonance, frequency alignment, and consciousness coherence, practical applications could enhance individual and collective well-being. This chapter explores how resonance principles might be applied in fields such as mental health, physical wellness, and social well-being. Through frequency-based devices, consciousness resonance technologies, and collective resonance practices, Conscienics has the potential to transform approaches to health, therapy, and community development.

1. Frequency-Based Health Technologies: Harnessing Resonance for Wellness

One promising application of Conscienics is in developing technologies that use specific frequencies to promote health and relaxation. Since the model suggests that consciousness and emotions operate at distinct frequencies, devices could help individuals tune into resonant frequencies that support well-being, relaxation, and mental clarity.

Potential Applications:

- **Binaural Beats**: Binaural beats, which use slightly different tones in each ear to

create a frequency associated with relaxation or focus, could be refined to align with specific resonant frequencies that promote emotional balance.
- **Pulsed Electromagnetic Field (PEMF) Therapy**: Conscienics-inspired PEMF devices could target frequencies that encourage physical and mental coherence, enhancing alignment within consciousness fields to promote overall health.
- **Vibrational Healing Devices**: Devices emitting frequencies through sound or light could be developed to promote harmony between mind and body, supporting a balanced state conducive to healing and well-being.

Key Insight: Frequency-based health technologies could harness resonance to help individuals achieve mental and physical wellness, creating alignment within consciousness fields for optimal well-being.

2. Consciousness Resonance Devices: Enabling Deeper Connections

Devices that facilitate consciousness resonance could enhance connections between individuals, allowing them to synchronize their mental states for improved empathy and collective focus. Such technology could support group meditation, collaboration, or shared intentions by helping people align within a shared consciousness field.

Potential Applications:

- **Group Meditation Technology**: Devices that monitor and synchronize participants' brainwaves or heart coherence during group sessions could facilitate resonance, enhancing shared intentions.
- **Virtual Reality (VR) Resonance Systems**: VR systems could simulate resonance experiences, enabling users to "tune in" to each other's consciousness

fields in real time, fostering collective empathy.

- **Remote Coherence Monitors**: Biofeedback devices that monitor heart-brain coherence remotely could allow people to connect and synchronize across distances, building virtual resonance bridges.

Key Insight: Consciousness resonance devices could bring people together by fostering shared mental states that enhance empathy, collaboration, and unity.

3. Resonance-Based Therapy: Healing Through Shared Intention

Therapeutic practices inspired by Conscienics could leverage resonance and shared focus to support mental and physical healing. These therapies would use group dynamics and collective intentions to help individuals achieve emotional and physical well-being.

Potential Applications:

- **Group Intention Therapy**: Similar to group meditation, this therapy would involve participants focusing on a shared intention for healing, creating a collective resonance that supports individual and group outcomes.
- **Biofield Therapies**: Techniques such as Reiki or Healing Touch could be enhanced by applying resonance principles to align consciousness fields, fostering harmony that promotes healing.
- **Cognitive Resonance Therapy**: A therapeutic approach where clients resonate with positive emotions, fostering mental clarity and emotional stability through shared focus with the therapist.

Key Insight: Resonance-based therapies could tap into the healing potential of collective focus, supporting therapeutic experiences that foster connection and well-being.

4. Sound and Light Therapy: Stimulating Consciousness Fields

Sound and light therapies use specific frequencies to promote healing and relaxation. In Conscienics, these therapies could target frequencies that harmonize with consciousness fields, creating resonance that supports mental and physical health.

Potential Applications:

- **Sound Baths**: Participants could experience resonant frequencies from instruments like singing bowls, aligning with consciousness field frequencies associated with peace or focus.
- **Color and Light Therapy**: Light frequencies could help patients achieve states of clarity and relaxation, using colors associated with particular mental and emotional states.

Key Insight: Sound and light therapies that target consciousness field frequencies could promote harmony within consciousness fields, supporting healing and balance.

5. Biofeedback for Resonance Training: Mastering Conscious States

Biofeedback technologies monitor physiological states like heart rate, brainwaves, or breathing, allowing individuals to observe their resonance state and learn self-regulation techniques. In Conscienics, biofeedback could help individuals achieve coherence within consciousness fields, fostering mental and emotional stability.

Potential Applications:

- **Resonance Training Programs**: Biofeedback devices could guide users to align their consciousness fields, helping them reach states of coherence that support emotional balance.
- **Collective Biofeedback Sessions**: In group settings, biofeedback could promote collective resonance, helping individuals experience group coherence and connection.
- **Self-Regulation for Emotional Balance**:

Biofeedback training could help individuals achieve resonant states, reducing stress and improving focus and resilience.

Key Insight: Biofeedback for resonance training could empower individuals to tune their consciousness fields, achieving balanced states that promote personal and collective well-being.

6. Community-Based Resonance Practices

Beyond individual applications, Conscienics could inspire community activities focused on shared resonance, such as group meditations, mindfulness sessions, or intention circles. These practices could encourage collective coherence, fostering social harmony and well-being.

Potential Applications:

- **Public Group Meditation Events**: Community-wide meditation sessions could encourage collective

resonance, reducing stress and building social harmony.
- **Community Intention Circles**: Regular community intention gatherings could amplify positive consciousness fields, creating a collective ripple effect that enhances community well-being.
- **Educational Programs**: Programs in schools and communities could teach resonance and consciousness field practices, fostering mindfulness and empathy across generations.

Key Insight: Community-based resonance practices could promote collective well-being, using shared intentions to enhance positive consciousness fields that benefit individuals and communities.

Summary: Bringing Conscienics into Practical Use

These applications illustrate how Conscienics could transform personal and community well-being through technologies, therapies, and community practices that enhance mental, emotional, and physical health. By harnessing resonance and collective focus, Conscienics-inspired methods could create positive consciousness fields that foster holistic wellness and social harmony.

In the following chapter, we will explore ethical considerations and guidelines for applying Conscienics in practical and clinical settings, ensuring responsible development and use of consciousness-based tools and practices.

Chapter 8: Philosophical and Ethical Implications of Conscienics

Introduction

The Conscienics model proposes that we are interconnected within a shared field of consciousness, suggesting that our thoughts, emotions, and intentions may influence others and even physical reality. While this model opens exciting possibilities, it also presents profound philosophical and ethical questions. If our minds are truly connected through resonance bridges and collective consciousness fields, what does this mean for personal responsibility, freedom, and the impact of our intentions on others?

This chapter explores the philosophical implications of Conscienics, addressing questions about individuality, free will, and the potential effects of collective intention. We also examine ethical challenges, such as the responsibility we may bear for our thoughts and emotions, the societal impact of collective resonance, and the risks of influencing others' consciousness fields. By confronting these issues, Conscienics becomes not only a model for understanding consciousness but also a framework for ethical reflection.

1. Personal Responsibility in a Shared Consciousness Field

If thoughts and emotions can influence others within a shared consciousness field, Conscienics suggests we have a responsibility for the energy we contribute to this field. Similar to our responsibility for physical actions, we might also consider how our mental states—

whether positive or negative—affect those around us through resonance.

This challenges the notion of a purely private inner life, proposing instead that our intentions and feelings may impact others in subtle yet meaningful ways. In Conscienics, this implies a need for emotional and mental coherence not only for individual well-being but also as an ethical commitment to others.

Key Insight: Conscienics reframes personal responsibility by suggesting that maintaining positive mental and emotional states is not solely self-serving but an ethical duty toward the collective.

2. Collective Influence and the Boundaries of Individual Autonomy

The Conscienics model, by positing that collective focus or intention can have tangible effects on individuals within the consciousness field, raises ethical questions about autonomy and collective influence. If a group's focused intention can influence others, it challenges the idea of individual autonomy by suggesting that thoughts and feelings are subject to collective resonance.

For example, large-scale collective intentions or meditative practices could positively impact participants and surrounding areas. However, what happens if these influences are directed toward unintended or invasive purposes? Conscienics calls for a careful balance between collective influence and respect for individual autonomy, ensuring that each person's freedom within the consciousness field is respected.

Key Insight: Conscienics urges us to find ethical boundaries for collective influence, protecting individual autonomy while exploring the power of collective resonance.

3. The Ethics of Using Resonance for Influence and Persuasion

If resonance bridges permit shared mental states and influence across consciousness fields, Conscienics could be used as a tool for persuasion. This potential brings ethical concerns about intentional influence and the manipulation of consciousness fields.

Even with positive intentions, using resonance principles to influence others' mental states must be carefully considered to avoid infringing on personal freedom.

In therapeutic, educational, or community settings, leaders or therapists could theoretically influence group consciousness fields. While the potential for support and healing is significant, Conscienics stresses the need for transparency and consent, respecting individuals' freedom to think and feel independently.

Key Insight: Conscienics supports the ethical application of resonance for influence, emphasizing the importance of transparency and respect for individual autonomy.

4. Free Will and Interconnected Consciousness

If consciousness is part of a shared, interconnected field, Conscienics brings free will into question. If our thoughts and emotions resonate with those of others, are we fully responsible for our individual states of mind? Are our choices and feelings shaped, even subtly, by the consciousness fields around us?

Conscienics suggests that while shared consciousness may influence us, we retain agency and choice within this field. This model encourages reflection on interconnectedness, urging a conscious approach to thoughts and actions, and fostering awareness of influences received and transmitted.

Key Insight: Conscienics doesn't eliminate free will but suggests that our choices and mental states are interconnected with others, encouraging mindful awareness within a shared field.

5. The Moral Responsibility of Collective Intentions

Conscienics posits that collective intentions can influence events, environments, and even physical systems. While group focus can promote positive outcomes, such as peace or healing, the ability to influence reality through shared intentions brings serious moral responsibility.

If collective intentions can create real-world effects, groups must approach this power with caution, aiming for constructive, inclusive, and respectful goals. By acknowledging the impact of shared intentions, Conscienics calls for ethical stewardship in guiding group focus toward compassionate outcomes.

Key Insight: Collective intentions are powerful tools within Conscienics, and groups should exercise this influence responsibly, focusing on intentions that promote well-being and inclusion.

6. Environmental Ethics and the Influence of Consciousness on Nature

Conscienics raises questions about environmental ethics if consciousness fields can interact with physical reality. Studies like Emoto's water crystal experiments and group intention research suggest that thoughts and emotions may impact natural systems, from plant growth to atmospheric conditions.

This potential influence encourages an ethical approach to nature, fostering conscious and respectful interactions with the environment. Through mindful practices, intention-setting, or collective focus, Conscienics offers a framework for environmental ethics as an extension of consciousness ethics, proposing that mental states contribute to natural harmony.

Key Insight: Conscienics emphasizes an ethical relationship with nature, acknowledging that thoughts and intentions may subtly impact environmental systems.

7. Integration with Spiritual and Existential Philosophy

A shared consciousness field aligns Conscienics with spiritual and existential philosophies that emphasize unity, non-duality, and collective purpose. Eastern traditions like Buddhism and Hinduism suggest interconnectedness, while Western existentialists explore individuality and shared experience. Conscienics bridges these perspectives, offering a scientifically informed model resonant with spiritual ideas of interconnectedness.

Incorporating spiritual and existential philosophy into Conscienics offers a broader context for understanding consciousness as both an individual and collective phenomenon. This framework encourages individuals to view life as part of a larger consciousness field, fostering empathy, connection, and shared purpose.

Key Insight: Conscienics harmonizes with spiritual and existential philosophy, promoting a view of consciousness as a collective experience that enhances empathy and unity.

Summary: Embracing Ethical Responsibility in the Conscienics Framework

The ethical and philosophical questions raised by Conscienics prompt reflection on how we interact with others and the world. If we are part of a shared consciousness field, we have a responsibility to approach our thoughts, emotions, and intentions with care, acknowledging their potential impact on others and the environment. Conscienics calls for an ethical framework that respects autonomy, collective responsibility, and the influence of our mental states on reality.

In the following chapter, we'll explore how Conscienics can be applied across disciplines, from psychology and sociology to education and environmental science. By extending the model beyond individual applications, we discover new ways to integrate Conscienics principles into diverse fields, enriching our understanding of consciousness and its role in society.

Chapter 9: Applications of Conscienics Across Disciplines

Introduction

Conscienics, as a unifying model of consciousness, holds the potential to influence multiple disciplines by proposing that consciousness operates as a resonant, interconnected field.

This perspective opens doors for innovative applications across fields such as psychology, sociology, education, environmental science, healthcare, conflict resolution, and technology. By integrating Conscienics principles into these areas, we gain new tools and frameworks that can enhance individual well-being, foster social cohesion, and promote a deeper connection with our environment. This chapter explores how Conscienics principles could transform these fields, inspiring interdisciplinary research and practical applications for societal benefit.

1. Psychology: Enhancing Mental Health and Therapeutic Approaches

Conscienics introduces tools and frameworks that can enrich mental health practices by incorporating resonance and collective consciousness. In therapeutic settings, understanding consciousness as a field allows for exploration into how shared mental states and resonance can improve well-being and foster resilience.

Applications in Psychology:

- **Resonance-Based Therapy**: Therapists can use resonance techniques to help clients synchronize with positive emotional states, fostering mental clarity and stress reduction.
- **Mindfulness and Biofeedback**: Conscienics-based mindfulness and biofeedback practices can teach clients to monitor and regulate their resonance states, achieving mental and emotional coherence.
- **Collective Therapy**: Group therapy sessions could focus on building collective resonance, creating a supportive environment where clients experience a deep sense of connection, accelerating healing.

Key Insight: Conscienics offers psychology new ways to foster mental well-being through resonance and collective consciousness.

2. Sociology: Understanding Collective Behavior and Social Cohesion

In sociology, Conscienics can deepen our understanding of collective behavior by providing a framework for how shared consciousness influences social dynamics. By examining resonance bridges and consciousness fields, sociologists can explore how collective thoughts, emotions, and goals shape societal behavior.

Applications in Sociology:

- **Collective Coherence Studies**: Studies can analyze how resonance within groups influences social harmony and cohesion during events or public gatherings.
- **Social Movements and Resonance**: Conscienics offers insights into how resonance bridges facilitate the spread of social movements and alignment of shared goals.
- **Community Healing Practices**: Community-based initiatives inspired by

Conscienics, such as group meditation and intention circles, can enhance collective well-being.

Key Insight: Conscienics provides a sociological lens for understanding collective resonance, fostering cohesive communities.

3. Education: Promoting Cognitive and Emotional Resonance in Learning

In education, Conscienics can help create environments that support collaborative learning, cognitive resonance, and emotional coherence. Integrating Conscienics principles into classrooms could enhance students' cognitive performance, emotional well-being, and teamwork skills.

Applications in Education:

- **Cognitive Resonance in Classrooms**: Activities can be designed to encourage shared focus and understanding, reinforcing group learning.
- **Emotional Coherence Practices**: Mindfulness and

reflective exercises can support emotional regulation, reducing stress and enhancing focus.
- **Holistic Curriculum Development**: A curriculum that includes cognitive and emotional coherence practices can foster lifelong well-being and social responsibility.

Key Insight: Conscienics can transform educational models by promoting supportive, resonance-based learning environments.

4. Environmental Science: Connecting Consciousness with Environmental Impact

Conscienics introduces a consciousness-based approach to environmental science, suggesting that human intention can influence natural systems. This model encourages mindful, resonant interactions with the environment, fostering ecological balance and awareness.

Applications in Environmental Science:

- **Collective Environmental Intentions**: Community events focused on ecological healing can promote collective resonance with nature, supporting environmental restoration.
- **Consciousness-Environment Interaction Research**: Studies on consciousness's influence on nature can build upon findings like Dr. Emoto's water research, exploring consciousness's potential impact on ecosystems.
- **Mindful Conservation Practices**: Conscienics-based conservation practices encourage mindful interactions with nature, promoting sustainable, resonant approaches.

Key Insight: Conscienics could inspire a more integrative approach to environmental science, promoting ecological health through resonance and mindful conservation.

5. Healthcare and Wellness: Resonance-Based Health Practices

In healthcare, Conscienics suggests that resonance-based practices could improve mental and physical health. By incorporating resonance techniques, healthcare providers can offer innovative approaches for preventive care and holistic wellness.

Applications in Healthcare:

- **Preventive Wellness Programs**: Programs that teach resonance and coherence techniques could enhance emotional resilience, stress management, and overall health.
- **Community Health Initiatives**: Group mindfulness and meditation events can promote community-wide coherence, supporting collective well-being.
- **Mind-Body Integration Techniques**: Integrating Conscienics into treatment plans encourages resonance between mental and physical states, fostering recovery and wellness.

Key Insight: Conscienics introduces resonance-based approaches that support holistic health, empowering individuals and communities.

6. Conflict Resolution and Peace Studies: Fostering Resonance and Empathy

Conscienics principles can advance conflict resolution by promoting shared resonance and collective empathy, supporting understanding and reducing tensions.

Applications in Conflict Resolution:

- **Resonance-Focused Mediation**: Mediators can use group resonance exercises to foster empathy and reduce hostility in conflict settings.
- **Collective Empathy Programs**: Programs that teach resonance techniques can help build understanding across cultural and ideological divides.
- **Global Peace Initiatives**: Large-scale resonance events, such as global meditations, can foster peace, creating a

consciousness field aimed at empathy and harmony.

Key Insight: Conscienics offers new pathways for peacebuilding and empathy, contributing to societal and global harmony.

7. Technology and Artificial Intelligence: Consciousness-Driven Innovation

As technology and AI evolve, Conscienics could inspire developments that align with human consciousness, promoting ethical and supportive interactions.

Applications in Technology and AI:

- **Resonant AI Systems**: AI systems designed to resonate with human consciousness fields could promote supportive interactions and well-being.
- **Consciousness-Coherent Virtual Spaces**: VR environments could use resonance technology to create immersive, collective resonance experiences.

- **Ethical AI Development**: Conscienics principles encourage AI developers to create systems that respect human values and autonomy, supporting mental and emotional health.

Key Insight: Conscienics could guide consciousness-aligned technological innovation, fostering ethical and supportive AI interactions.

Summary: Transforming Disciplines with Conscienics Principles

The applications outlined in this chapter illustrate how Conscienics could enhance diverse fields, bridging science, social responsibility, and holistic wellness. By promoting resonance, collective consciousness, and interconnectedness, Conscienics offers a transformative framework that can inspire new methods, tools, and perspectives for fostering personal and societal well-being.

In the next chapter, we will examine the roadmap for advancing Conscienics through empirical research, interdisciplinary collaborations, and educational initiatives, setting the stage for Conscienics to become a scientifically grounded, impactful model for understanding and fostering consciousness.

Chapter 10: Environmental Ethics and Influence of Consciousness on Nature

Introduction

The Conscienics model introduces a profound ethical framework for our relationship with the environment. By suggesting that consciousness operates as a field capable of influencing physical systems, including ecosystems, Conscienics challenges us to consider the impact of our thoughts and intentions on the natural world. If our mental states, emotions, and collective focus can interact with and even shape environmental systems, we bear a moral responsibility to approach nature with mindfulness, respect, and intention. This chapter explores the environmental implications of Conscienics, considering how conscious practices could promote ecological balance and encouraging a framework of environmental ethics grounded in resonance and collective consciousness.

1. Consciousness and Environmental Interaction: Insights from Research

Research into consciousness and environmental interaction, such as Dr. Masaru Emoto's studies on water crystallization, suggests that human thoughts and emotions can influence natural systems. Although Emoto's work is still debated in scientific circles, it has inspired further inquiry into how collective focus or intention might impact plants, water, and even atmospheric conditions. These studies suggest that environmental influence through consciousness may be subtle yet measurable, raising the possibility that collective resonance practices could support ecological balance.

Research Highlights:

- **Emoto's Water Crystallization**: Emoto's experiments found that water

exposed to positive intentions formed more harmonious crystal structures, suggesting that thoughts may have a physical impact on water molecules.
- **Global Consciousness Project**: Studies measuring random number deviations during global focus events indicate that collective mental states may influence physical systems, offering insights into how collective intention might affect environmental factors.

Key Insight: Conscienics supports the idea that consciousness can interact with environmental systems, suggesting that human intention may contribute to ecological health through resonance-based practices.

2. Environmental Mindfulness: An Ethical Approach to Nature

In Conscienics, environmental ethics extend beyond action to encompass thought and intention. By encouraging people to practice environmental mindfulness, Conscienics promotes an ethical approach that includes respecting natural systems and harmonizing with the environment through conscious resonance. Whether through meditation, intention-setting, or eco-mindfulness, individuals can align their consciousness fields with nature, contributing positively to the environment.

Suggested Practices:

- **Eco-Meditation**: Practicing meditation with a focus on natural systems, aiming to resonate with the rhythms and vitality of the environment.
- **Intention-Setting for Nature**: Collective gatherings where participants focus on positive intentions for ecological balance, biodiversity, or environmental restoration.
- **Mindful Conservation**: Practicing conservation efforts with intentional mindfulness, fostering a respectful relationship with ecosystems that supports natural harmony.

Key Insight: Environmental mindfulness practices in Conscienics encourage a respectful and resonant approach to interacting with nature, fostering an ethical commitment to environmental harmony.

3. Group Intention and Ecological Impact: The Potential for Collective Resonance

If collective resonance has measurable effects on physical systems, as proposed by Conscienics, group intention could be directed toward environmental goals, such as conservation, climate stabilization, or ecosystem restoration. Large-scale intention gatherings, similar to global meditation events, could focus on ecological outcomes, aiming to amplify collective resonance and contribute positively to the environment.

Application Examples:

- **Community Environmental Healing Events**: Communities could organize events where participants meditate or focus on environmental goals, creating a collective consciousness field dedicated to ecological well-being.

- **Global Environmental Intention Days**: International events could bring people together to meditate or set intentions for planetary health, potentially supporting positive shifts in ecological systems.
- **School and Community Initiatives**: Educational programs could teach students about the environmental impact of conscious intention, encouraging mindfulness toward nature and fostering ecological responsibility.

Key Insight: Conscienics proposes that collective resonance for environmental goals could create positive consciousness fields that support ecological health, suggesting a proactive role for consciousness in environmental conservation.

4. Ethical Stewardship and Environmental Responsibility

Conscienics introduces a framework of ethical stewardship for our relationship with the environment, where the impact of thoughts, emotions, and collective intentions on nature is acknowledged. This approach suggests that, just as we bear responsibility for our physical actions, we also carry ethical obligations regarding our mental states and intentions toward the environment.

Principles of Ethical Stewardship:

- **Respect for Nature's Resonance**: Recognizing that natural systems may resonate with human consciousness fields, we should approach the environment with a mindset of respect and harmony.
- **Mindful Intention-Setting**: When engaging in intention-setting for environmental goals, individuals and groups should commit to ethical considerations, ensuring intentions are positive, inclusive, and respectful of all life forms.
- **Environmental Accountability**: Conscienics suggests that individuals reflect on the potential environmental impact of their thoughts and emotions, fostering an inner sense of

responsibility toward the planet.

Key Insight: Ethical stewardship in Conscienics calls for conscious, respectful interactions with nature, acknowledging that our consciousness fields may influence environmental systems.

5. Integrating Conscienics into Environmental Science and Policy

The Conscienics model suggests that integrating consciousness fields into environmental science and policy could enhance ecological protection efforts. By combining traditional conservation approaches with consciousness-based practices, environmental policy could promote ecological balance through both action and mindful intention. Policies that encourage community mindfulness events, collective environmental intentions, and conservation education can harness the principles of Conscienics for greater environmental stewardship.

Potential Policy Applications:

- **Mindful Conservation Policies**: Policies promoting eco-mindfulness and community environmental intention events could support a holistic approach to ecological conservation.
- **Environmental Education Programs**: Integrating Conscienics principles into educational curricula could foster eco-consciousness, teaching future generations the value of intentional, respectful interaction with nature.
- **Research Initiatives**: Government and private organizations could fund research into consciousness-environment interactions, exploring the practical implications of resonance and intention on ecological health.

Key Insight: Integrating Conscienics into environmental science and policy offers a pathway for holistic ecological conservation, supporting mindful practices that promote both environmental and consciousness well-being.

Conclusion: Toward a Conscious Environmental Ethic

The Conscienics model encourages a reimagined relationship with the environment—one that acknowledges the potential influence of human consciousness on ecological systems. By fostering mindfulness, collective resonance, and respectful intention-setting, Conscienics suggests that we can contribute positively to the planet's health. This approach combines science and ethics, challenging us to be mindful stewards of both our mental states and the natural world. Through environmental ethics grounded in resonance, Conscienics proposes a vision for a sustainable future where individuals and communities can harmonize with nature, creating a healthier, more connected planet.

In the next chapter, we'll explore the broader implications of Conscienics across diverse fields, such as psychology, sociology, education, and technology, revealing the transformative potential of consciousness fields in modern society.

Chapter 11: Social Applications and Community Practices in Conscienics

Introduction

Conscienics introduces a vision of interconnected consciousness fields, where individuals and groups can influence collective well-being through resonance, empathy, and shared intentions. This model lends itself to practical applications in community building, social cohesion, and collective healing. By implementing Conscienics principles in community practices, we can foster environments that support personal growth, empathy, and resilience, ultimately strengthening social bonds. This chapter explores how Conscienics can enhance community experiences, create resonance-focused social programs, and empower individuals to contribute positively to collective consciousness fields.

1. Community Meditation and Mindfulness Programs

One of the most accessible ways to integrate Conscienics principles into community life is through group meditation and mindfulness programs. These practices can encourage coherence within community consciousness fields, enhancing empathy, reducing stress, and fostering a shared sense of well-being.

Program Suggestions:

- **Weekly Community Meditation Sessions**: Organized meditation events allow community members to gather regularly, aligning their mental and emotional states to create collective coherence and mutual support.
- **Mindfulness and Resonance Workshops**: Offering workshops on resonance and mindfulness can teach individuals how to cultivate positive states that contribute to community well-being.
- **School-Based Mindfulness Programs**: Schools can integrate mindfulness practices into their curriculum, helping students learn emotional regulation and resilience, and fostering a culture of empathy from a young age.

Key Insight: Community meditation and mindfulness programs promote resonance and emotional alignment, strengthening community ties and fostering a shared consciousness field that supports mutual well-being.

2. Intention Circles and Collective Resonance Events

Intention circles are gatherings where participants focus on a shared goal, such as healing, peace, or personal growth. By aligning mental and emotional states, intention circles can amplify resonance within community consciousness fields, creating positive energy that extends beyond the immediate group.

Application Ideas:

- **Healing Intention Circles**: Communities can hold regular circles where participants focus on collective healing, supporting each other through resonance and shared compassion.
- **Seasonal or Holiday Intention Events**: Holding intention circles during

important cultural events or seasonal transitions can unite communities around shared goals, enhancing collective harmony.
- **Youth and Family Intention Circles**: Family-focused or youth intention circles promote resonance within family units, supporting positive relationships and intergenerational understanding.

Key Insight: Intention circles allow communities to channel collective resonance toward meaningful goals, promoting healing, unity, and positive social impact.

3. Empathy Training and Resonance-Building Exercises

Empathy training, rooted in Conscienics principles, emphasizes the ability to tune into others' emotions and perspectives. By practicing empathy and resonance-building exercises, individuals can create resonance bridges, promoting social harmony and understanding.

Program Suggestions:

- **Empathy Workshops for Schools and Workplaces**: Teaching empathy-building exercises within educational or workplace settings fosters a culture of respect and understanding, reducing conflict and promoting collaboration.
- **Community Compassion Circles**: Regular compassion circles allow participants to practice empathic listening and mutual support, building resonance within the community.
- **Conflict Resolution Training with Resonance Principles**: Introducing resonance principles into conflict resolution can help individuals approach disagreements with empathy, enhancing social harmony.

Key Insight: Empathy training grounded in Conscienics principles fosters resonance and understanding within communities, promoting unity and reducing interpersonal conflicts.

4. Community Healing Events and Resonance Practices

Community healing events invite people to gather with the shared goal of promoting physical, mental, and emotional well-being for themselves and others. By aligning intentions and practicing resonance-based activities, such as group meditation, sound baths, or intention setting, these events enhance the consciousness field of the community, supporting collective health and harmony.

Event Ideas:

- **Sound Healing Gatherings**: Community sound healing events, featuring instruments like singing bowls or gongs, create a resonant environment that supports relaxation and unity within the community.
- **Healing Walks and Nature Retreats**: Organized walks or retreats in nature allow participants to connect with the environment, promoting both personal and community well-being through

mindfulness and resonance with natural surroundings.
- **Cultural Healing Days**: Community events honoring cultural or seasonal practices can strengthen community identity, fostering shared resonance and mutual respect.

Key Insight: Community healing events amplify collective resonance, fostering well-being and promoting a strong, supportive consciousness field within communities.

5. Educational Programs on Consciousness and Resonance

Educational initiatives that introduce Conscienics principles to schools, universities, and community centers can empower individuals to understand and apply resonance in their daily lives. By learning about consciousness fields, resonance, and collective coherence, individuals can become more aware of their impact on others and the community.

Educational Program Ideas:

- **Consciousness Studies in Schools**: Introducing basic concepts of Conscienics, resonance, and collective consciousness can encourage students to develop empathy, mindfulness, and a sense of shared responsibility.
- **Workshops for Community Leaders**: Providing training for community leaders on resonance and collective well-being can equip them with tools to foster a positive, supportive environment.
- **Public Lectures and Webinars**: Offering accessible lectures and webinars on Conscienics principles allows the wider public to engage with the model, fostering understanding and enthusiasm for community practices.

Key Insight: Education on Conscienics principles equips individuals and leaders with the tools to contribute positively to community resonance, fostering a sense of shared consciousness and social responsibility.

6. Integrating Conscienics into Social Policy and Community Planning

Social policies and community planning efforts that incorporate Conscienics principles can foster environments that support resonance, collective coherence, and mental and emotional well-being. By designing public spaces, social programs, and community resources with collective resonance in mind, cities and towns can create environments that enhance social cohesion and community resilience.

Policy and Planning Suggestions:

- **Resonance-Friendly Public Spaces**: Design parks and community centers that promote calm, relaxation, and collective resonance through natural elements, meditation spaces, and sound-absorbing materials.
- **Social Support Programs**: Policies that fund programs for collective meditation, intention circles, and empathy training in underserved communities

can foster resilience, support mental health, and reduce social isolation.
- **Resonance-Based Community Events**: Municipalities can sponsor events focused on mindfulness, community coherence, and positive resonance, promoting social harmony and unity.

Key Insight: Integrating Conscienics principles into social policy and community planning fosters inclusive, resonance-supportive environments that promote collective well-being.

Summary: Strengthening Communities Through Conscienics Principles

The applications and practices outlined in this chapter show how Conscienics can foster community connection, empathy, and collective resilience. By integrating resonance-based practices into social programs, educational initiatives, and public policy, communities can create supportive consciousness fields that enhance individual and collective well-being. Conscienics offers a transformative approach to community-building, inspiring practices that strengthen social bonds and create environments where individuals can thrive together.

In the next chapter, we will examine future directions for Conscienics research, proposing a roadmap for scientific inquiry, interdisciplinary collaboration, and community engagement that can further develop the model and its applications.

Chapter 12: Future Directions and Research Agenda

Introduction

The Conscienics model represents a groundbreaking approach to understanding consciousness, proposing that consciousness operates as a resonant field that connects individuals and potentially influences physical reality. For Conscienics to transition from a theoretical model to an empirically validated framework, a well-structured research agenda is essential. This chapter outlines a roadmap for advancing Conscienics through experimental validation, methodological development, interdisciplinary collaboration, and public engagement. By pursuing these directions, we can lay a foundation for Conscienics as a respected model in scientific, social, and practical domains.

1. Experimental Validation of Consciousness Fields and Resonance

Empirical validation is a primary goal for establishing Conscienics as a credible framework. Initial studies should aim to test the existence and influence of consciousness fields, resonance bridges, and collective coherence effects.

Research Goals:

- **Non-Local Resonance Testing**: Conduct studies measuring physiological coherence (e.g., brainwave and heart rate synchronization) among individuals in shared or remote locations to investigate non-local resonance.
- **Collective Intention Effects**: Organize large-scale

experiments involving group meditations or focused intention sessions to observe potential impacts on physical environments, such as atmospheric changes or deviations in random event generators.
- **Microtubule and Quantum Coherence**: Investigate the role of microtubules in consciousness by studying their behavior during altered states, such as meditation, using imaging techniques to assess any coherence patterns linked to non-local effects.

Key Outcome: Experimental validation of resonance and consciousness fields will provide a scientific basis for Conscienics, establishing credibility and inspiring further exploration.

2. Development of Methodological Standards and Measurement Tools

Creating reliable methodologies and tools for studying consciousness fields is essential. This involves developing technologies that can measure resonance, coherence, and non-local effects accurately.

Research Goals:

- **Specialized Biofeedback Devices**: Develop biofeedback instruments that can measure resonance and coherence within consciousness fields, such as EEG and HRV tools tailored for detecting synchronization across individuals.
- **Consciousness Field Detection Instruments**: Invest in developing instruments that can detect shifts in resonance and coherence within groups, capturing the presence of consciousness fields.
- **Standardized Protocols**: Establish research protocols that standardize experimental designs, data collection, and analysis for Conscienics studies.

Key Outcome: New tools and standards will support consistent, replicable research on Conscienics, building a foundation of credible studies and advancing the model's acceptance.

3. Establishing Interdisciplinary Research Collaborations

The Conscienics model invites collaboration across fields, including neuroscience, psychology, physics, and sociology. By integrating diverse perspectives, interdisciplinary research can provide a comprehensive exploration of Conscienics principles.

Research Goals:

- **Neuroscience and Quantum Physics**: Encourage partnerships to explore possible quantum processes in the brain related to non-local consciousness.
- **Social Psychology and Group Dynamics**: Collaborate with social psychologists to study group resonance and the influence of collective focus on behavior and mental states.
- **Environmental Science and Consciousness**: Partner with

environmental scientists to study consciousness-environment interactions, exploring how collective intentions may impact ecological systems.

Key Outcome: Interdisciplinary collaborations will expand Conscienics' research scope and increase its real-world applications, fostering innovation across scientific and social fields.

4. Practical Applications in Health and Well-Being

Conscienics offers promising applications in health and wellness, particularly through resonance-based therapies, collective coherence practices, and preventive programs.

Research Goals:

- **Clinical Trials on Resonance-Based Therapies**: Conduct trials on therapies that use resonance to improve mental health outcomes, including stress reduction,

emotional resilience, and cognitive clarity.
- **Community Health Programs**: Implement group coherence initiatives in community health settings, studying the impact of collective meditation on public health indicators like crime and stress.
- **Preventive Wellness Programs**: Develop programs that teach resonance techniques for emotional balance and resilience as preventive health measures.

Key Outcome: Practical health applications will demonstrate the benefits of Conscienics, making its principles accessible to individuals and communities alike.

5. Educational Programs and Public Awareness Initiatives

Public engagement is crucial for Conscienics to reach its full potential. Educational programs and awareness initiatives can introduce resonance and coherence concepts to a wider audience, empowering individuals to apply these principles.

Research Goals:

- **Educational Curricula Development**: Design curricula for schools, universities, and community centers, teaching principles of collective consciousness, resonance, and ethical responsibility.
- **Workshops and Seminars**: Host public events that teach Conscienics practices for achieving coherence and cultivating positive intentions.
- **Collaboration with Wellness Organizations**: Partner with organizations to create programs that incorporate Conscienics, reaching a broader audience interested in mindfulness and personal growth.

Key Outcome: Public awareness initiatives will integrate Conscienics into mainstream culture, allowing people to apply its principles in daily life and fostering a broader understanding of collective consciousness.

6. Funding and Institutional Support for Conscienics Research

Sustained research on Conscienics requires dedicated funding and institutional backing. Establishing partnerships with research institutions, foundations, and government agencies is essential.

Research Goals:

- **Grant Applications for Innovative Research**: Apply for funding from foundations that support consciousness studies and interdisciplinary research.
- **Establish Conscienics Research Centers**: Create centers within universities or independent institutions to coordinate ongoing research and collaborations.
- **Engagement with Government and Private Sector**: Encourage partnerships with entities interested in wellness,

education, and mental health, promoting Conscienics as a tool for public health and social cohesion.

Key Outcome: Institutional support will ensure resources for long-term Conscienics research, advancing its development into a recognized field of study.

7. Publishing and Building a Research Community

To gain scientific acceptance, it is essential to publish findings and build a Conscienics research community. Academic publications and conferences will encourage knowledge sharing and collaborative exploration.

Research Goals:

- **Peer-Reviewed Publishing**: Publish studies on Conscienics in scientific journals to build a foundational body of literature.
- **Conferences and Symposia**: Host Conscienics-focused events where researchers can

share findings, exchange ideas, and foster collaborative projects.
- **Online Research Networks**: Develop a platform for researchers to connect, share insights, and coordinate research efforts, building a collaborative community.

Key Outcome: Publishing and community-building will create a solid academic base for Conscienics, fostering ongoing exploration and contributing to its growth as a respected model.

Summary: A Roadmap for the Future of Conscienics

This roadmap invites researchers, educators, and practitioners to advance Conscienics through experimental validation, interdisciplinary collaboration, and public engagement. By pursuing this research agenda, we can support Conscienics as a scientifically grounded framework that deepens our understanding of consciousness, connection, and collective power. Through this collaborative journey, Conscienics has the potential to transform how we understand and interact with consciousness, creating a model that bridges science, philosophy, and practical applications for personal and societal benefit.

In the next chapter, we'll explore the interplay of structure and experience in consciousness, discussing how Conscienics aligns with and expands upon foundational theories in the study of consciousness.

Chapter 13: The Interplay of Structure and Experience in Consciousness

Introduction: Bridging Two Perspectives of Consciousness

Consciousness has long intrigued scientists, philosophers, and spiritual thinkers, prompting numerous theories that aim to capture its elusive nature. In contemporary discussions, two perspectives often emerge: the **Unified Field Theory** of consciousness, which offers a structural and universal approach, and the **Conscienics model**, emphasizing consciousness as a participatory, experiential field. By exploring these theories together, we aim to offer a framework that bridges the stable foundation provided by the Unified Field Theory with the dynamic, experiential dimension offered by Conscienics. This chapter examines the interplay between these perspectives, illustrating how they collectively form a holistic understanding of consciousness.

Unified Field Theory: The Structural Base of Consciousness

The Unified Field Theory posits that consciousness is a foundational, omnipresent field of information in which all things are interconnected. Rooted in quantum mechanics and information theory, this perspective proposes that consciousness provides a stable informational matrix underlying reality itself. Nobel laureate Erwin Schrödinger once suggested that "consciousness is a singular of which the plural is unknown," alluding to a unity that underlies individual awareness. Unified Field Theory thus treats consciousness as a structural ground, an informational fabric upon which all reality rests.

Core Principles:

- **Consciousness as Information**: Unified Field Theory posits that all phenomena, including physical and mental processes, are expressions of a fundamental informational structure.
- **Non-Locality and Interconnectivity**: Drawing from quantum mechanics, this theory suggests that consciousness, much like entangled particles, can connect instantly across distances, forming an interconnected "fabric" that defies traditional spatial boundaries.

Scientific Underpinnings:

- **Quantum Mechanics**: The phenomenon of quantum entanglement, where particles remain connected despite separation, suggests that consciousness could similarly maintain instantaneous connections across the universe.

- **Information Theory**: Physicist John Wheeler's principle, "it from bit," aligns with the concept that consciousness is fundamentally informational, creating a vast network or field encompassing all processes, minds, and matter.

Conscienics: The Experiential and Participatory Dimension of Consciousness

While Unified Field Theory emphasizes structural connectivity, Conscienics views consciousness as a resonant and participatory field. In this model, consciousness is not static but actively co-created, shaped by the interactions and awarenesses of its participants. Conscienics presents consciousness as a vibrational, dynamic field where each conscious entity contributes to the shared resonance through unique thoughts, emotions, and intentions.

Core Principles:

- **Consciousness as a Resonant Field**: Each entity resonates

within a shared spectrum, suggesting that consciousness manifests as waves of interaction within a continually evolving field.
- **Dynamic Participation**: Conscienics suggests that each individual's experiences and awareness shape the collective consciousness field, creating an adaptive, co-created reality.

Experiential Aspects:

- **Vibrational Connectivity**: Conscienics proposes that emotions and thoughts resonate across consciousness fields, creating harmonized "frequencies" of connection.
- **Participatory Reality**: Similar to quantum observation, where the observer affects the experiment outcome, each person's awareness "tunes" the consciousness field, shaping its collective resonance.

Interdependence of Structure and Experience

The interplay between these theories provides a balanced framework, viewing consciousness as both structurally interconnected and experientially dynamic. The structural aspects of Unified Field Theory offer stability and cohesion, while Conscienics' participatory perspective brings vibrancy and adaptability. Together, they imply a consciousness that is both a foundational field and a living, evolving experience shaped by its participants.

Key Interactions:

- **Structural Cohesion**: Unified Field Theory grants consciousness a universal, cohesive substrate where all entities are embedded within a vast, interconnected matrix.
- **Experiential Adaptability**: Conscienics enriches this structure, positioning consciousness as a responsive, participatory field continually shaped by resonant interactions.

Dual Perspectives on Consciousness:

1. **Structural**: The Unified Field Theory emphasizes unity, where consciousness binds all entities within a singular informational framework.
2. **Participatory**: Conscienics introduces individuality within this unity, suggesting that consciousness is varied, alive, and co-creative, with each participant adding unique resonance to the field.

Applications and Implications of the Combined Model

The unified framework has profound implications, encouraging a balanced approach to personal and collective consciousness by recognizing both foundational and participatory aspects. It allows for practical applications in personal development, social cohesion, and even the bridging of science and spirituality.

1. **Personal and Collective Growth**:
   - **Personal Development**: Understanding consciousness as both a stable field and an interactive experience empowers individuals to enhance their resonance, aligning with the collective consciousness field through practices like meditation and intention-setting.
   - **Collective Consciousness and Social Change**: Recognizing that each person influences the collective consciousness field supports the potential for social practices, such as group meditation and synchronized intention, to elevate the collective state.
2. **Bridging Science and Spirituality**:

- The combined model harmonizes scientific rigor with spiritual insight, addressing the "hard problem" of consciousness by recognizing both an informational structure and subjective experience.
- It proposes a balanced understanding, embracing scientific measurement while valuing experiential knowledge, ultimately offering a pathway for interdisciplinary study of consciousness that respects intuition alongside empirical inquiry.

Future Research Directions

The interplay between Unified Field Theory and Conscienics opens exciting avenues for research, encouraging interdisciplinary exploration that combines neuroscience, psychology, physics, and consciousness studies.

Research Goals:

- **Empirical Studies on Resonance**: Investigate how individual consciousness interacts with the collective field through studies in biofield science, neuroplasticity, and psychophysiology.
- **Consciousness-Driven Technologies**: Develop technologies that harness resonance, non-locality, and collective consciousness, such as AI systems that augment empathy or tools that enhance alignment within the consciousness field.

Conclusion: Toward a Unified Theory of Conscious Experience

In summary, the interplay between the Unified Field Theory and Conscienics offers a nuanced view of consciousness as both universal and uniquely personal. This framework encourages readers to see themselves as both integral parts of a unified field and as active participants shaping that field. By embracing both structural and experiential aspects of consciousness, we move closer to understanding the full spectrum of existence, recognizing the interconnected reality that unites us all.

Chapter 14: Introduction to the Supplementary Manuscripts

Connecting Conscienics with Applied Frameworks

With the foundational principles and applications of Conscienics established, this final chapter introduces two supplementary manuscripts: **"A Unified Theory of Additive and Subtractive Color"** and **"Anthrocosmia: A Multilayered Framework Integrating Human Consciousness within a Cosmic Context."** Each manuscript offers a focused exploration of specific concepts within Conscienics, providing practical applications and metaphors that deepen our understanding of consciousness fields, resonance, and collective experience. Together, these manuscripts extend Conscienics principles into new territories, offering readers tools and frameworks to apply and visualize these ideas in diverse domains.

Manuscript 1: A Unified Theory of Additive and Subtractive Color

This manuscript explores the principles of color theory through additive (RGB) and subtractive (CMY) models, revealing parallels with the Conscienics model's emphasis on resonance, duality, and unity. Just as additive colors (light) and subtractive colors (pigment) combine or filter to create diverse color experiences, the consciousness field can be seen as a spectrum where individual frequencies either harmonize or create dissonance.

Connection to Conscienics:

- **Metaphor for Consciousness Fields**: Just as colors interact to produce new shades, consciousness fields interact to produce unique resonant experiences. The interaction of additive and subtractive color models provides a visual analogy for the different ways consciousness states may

align, blend, or create contrasts within a shared field.
- **Duality and Integration**: The manuscript uses the complementary nature of color models as a metaphor for the balance of individuality and unity within consciousness, illustrating how diverse elements harmonize within the Conscienics model.

This manuscript provides readers with a visually accessible way to understand complex ideas within Conscienics, using the interplay of color as a practical and symbolic tool for comprehending consciousness resonance.

Manuscript 2: Anthrocosmia—Integrating Human Consciousness with the Cosmos

"Anthrocosmia" expands Conscienics principles to a cosmic scale, presenting a multilayered framework that situates individual consciousness within a vast, interconnected cosmic context. By viewing consciousness as a field that spans personal, communal, and cosmic layers, this manuscript illustrates how resonance bridges can connect individual minds not only to one another but to larger universal patterns.

Connection to Conscienics:

- **Personal and Cosmic Integration**: This framework illustrates the idea that each consciousness field participates in a greater universal resonance, suggesting that individual intentions and awareness contribute to the larger cosmic field.
- **Multi-Layered Resonance**: "Anthrocosmia" positions consciousness as a dynamic field influenced by cosmic forces, emphasizing that our personal experiences and resonances may align with broader cosmic frequencies, supporting Conscienics' proposal of non-local, collective consciousness fields that extend across personal and cosmic scales.

Through "Anthrocosmia," readers can appreciate how Conscienics principles apply within a universal context, encouraging reflection on the interconnectedness of individual, social, and cosmic consciousness fields.

Conclusion: Expanding Conscienics Through Applied Exploration

The two manuscripts provide practical frameworks for understanding and applying Conscienics principles across multiple dimensions of experience. By introducing metaphors of color theory and cosmic alignment, these texts make complex concepts more accessible, offering readers tools to visualize and engage with the resonance principles that lie at the heart of Conscienics. Whether exploring consciousness through color or the cosmos, these manuscripts provide pathways to deeper engagement, helping readers integrate Conscienics into personal and universal contexts.

In this way, the Conscienics model evolves from a theoretical framework into a multidimensional tool, encouraging exploration, personal resonance, and a richer understanding of the shared consciousness field that connects us all.

Author's Notes: Reflections and Aspirations - A Vision for the Future of Conscienics

As I reflect on the journey that led to the creation of Conscienics, I am inspired by the potential this model holds—not just as a scientific framework but as a way to reimagine our connections to each other and the world around us. Conscienics was born from a desire to understand consciousness beyond the boundaries of the individual mind, exploring how thoughts, emotions, and intentions could connect us on a fundamental level. This journey has drawn from the work of great thinkers and emerging research, but it has also been deeply personal, driven by a belief that consciousness itself may be a bridge that can unite science, spirituality, and humanity.

A Call for Unity in an Interconnected World

At its core, Conscienics is a vision of unity. It suggests that our thoughts and feelings resonate beyond our individual selves, reaching out into a shared field that connects us all. In a world often divided by conflict and misunderstanding, this model offers a hopeful perspective, one where our collective focus and intentions can create harmony, empathy, and peace. If we are truly connected through a consciousness field, then each of us has the power to influence the whole, fostering positive change simply by cultivating coherence and compassion within ourselves.

The Dream of a New Paradigm

I envision a future where Conscienics becomes a cornerstone of a new paradigm—a world where science embraces the interconnected nature of consciousness and recognizes the profound impact our inner lives have on each other and on reality itself. Imagine if healthcare, education, and social systems were designed with an understanding of resonance and coherence, supporting people not only in body but in mind and spirit. Imagine if communities prioritized shared focus and collective well-being, knowing that each person's state of mind contributes to the whole.

Conscienics could inspire a movement toward a more conscious society, one where each individual's well-being is valued as part of the collective good.

A Vision for Future Generations

I hope Conscienics will inspire future generations to explore consciousness with open minds and innovative spirits. As more research and collaboration bring new insights, I imagine Conscienics evolving, incorporating discoveries from neuroscience, quantum mechanics, psychology, and beyond. Ultimately, I hope this model will encourage young thinkers, scientists, and explorers to approach consciousness with curiosity and reverence, recognizing it as a powerful, unifying force woven into the fabric of our lives.

An Invitation to Reflect and Contribute

The journey of Conscienics is just beginning, and its growth depends on the contributions of many voices, perspectives, and fields. I invite each reader—whether a scientist, a spiritual seeker, or simply someone curious about consciousness—to reflect on these ideas and consider how they might contribute to this vision. Whether through personal practice, research, teaching, or simply cultivating mindful awareness, each of us can play a role in bringing Conscienics to life.

In Closing: The Power of Connection

At its heart, Conscienics reminds us that we are not isolated beings but part of a greater whole. Each thought, each intention, each moment of kindness ripples outward, touching others in ways we may never see but that nonetheless shape our shared reality. If we choose to live with this awareness, cultivating positive resonance within ourselves and with others, we have the power to create a world where connection, compassion, and understanding are at the center of our lives.

A Unified Theory of Additive and Subtractive Color in Physical and Metaphysical Contexts: Expanding Perception Beyond Sight

Abstract

This paper introduces a unified theory of color perception, integrating additive and subtractive models within scientific, metaphysical, and multisensory contexts to explore color as both a physical and experiential phenomenon. In additive color mixing (RGB model), wavelengths of red, green, and blue light converge to create white, symbolizing integration and expansion. In contrast, subtractive color mixing (CMY model) with pigments absorbs light, producing black—a representation of material limitation. The symbolic implications of these models align with metaphysical principles, such as Hermetic duality and the concept of "as above, so below," suggesting that additive and subtractive color models represent contrasting states of awareness: unity versus sensory limitation.

Recent studies on non-visual color perception challenge the traditional assumption that color is exclusively a visual experience. Empirical research in sensory substitution and cross-modal perception reveals that color can be discerned through sound and vibration, potentially allowing humans to perceive color beyond sight alone. This paper explores applications in assistive technologies, multisensory art, and therapeutic practices, using Solfeggio frequencies to investigate possible correlations between sound and color. By examining the neuroplasticity involved in sensory adaptation, we propose that consciousness itself may operate through a multisensory, adaptable framework.

The findings support a model where color is both a physical reality and a symbolic bridge to understanding perception, suggesting that human awareness integrates sensory inputs dynamically. Future research in technology, art, and neuroscience can expand on these insights, contributing to fields such as cognitive science, immersive media, and consciousness studies.

Introduction

Color Perception and the Nature of Reality

The phenomenon of color has captivated human curiosity for centuries, serving as a subject of study in physics, art, philosophy, and now neuroscience. Color is not only a sensory experience but also a potential gateway into understanding the mind's capacity to interpret and adapt to the world. Two primary models of color perception—additive and subtractive—illustrate distinct ways in which color manifests physically and symbolically. In additive color mixing, used in light-based media, red, green, and blue wavelengths combine to form white light, symbolizing an expansion or synthesis of color. This RGB model has been fundamental in fields ranging from digital technology to visual arts, where light is used to create an array of colors that merge to form a sense of unity.

In contrast, subtractive color mixing, used with physical pigments and dyes, relies on the primary colors cyan, magenta, and yellow to absorb light and produce black, suggesting a principle of limitation or material grounding. The CMY model, central to traditional art forms and print media, reflects a process where colors result from the selective absorption and reflection of light, producing color through the absence of certain wavelengths. This duality between additive and subtractive mixing provides an intriguing lens for exploring how color functions as both an integrative and differentiating force. The convergence of light in additive mixing symbolizes unity and synthesis, while the absorptive nature of subtractive mixing aligns with sensory specificity and material limitation.

Symbolic and Metaphysical Interpretations of Color Models

Beyond their physical properties, these color models have rich metaphysical and symbolic implications, which suggest that color perception might serve as a bridge between material and transcendent realities. The Hermetic principle of "as above, so below" proposes that the physical and spiritual realms mirror one another. In this context, additive color mixing—where light converges to create white—represents unity, expansiveness, and interconnectedness, aligning with spiritual ideals of integration and harmony. Subtractive mixing, on the other hand, symbolizes sensory absorption, limitation, and individuality, suggesting that color embodies both the expansive potential of consciousness and its sensory boundaries.

Eastern philosophical traditions further enhance these symbolic interpretations. In Taoism, for example, the concept of yin and yang encapsulates a duality wherein complementary forces, such as light and darkness, interact to form a harmonious whole. This philosophy can be applied to color theory, where additive mixing aligns with yang (expansive, light-giving) and subtractive mixing with yin (absorptive, grounding). Hindu metaphysics similarly emphasizes a dual nature in existence, where Brahman, the ultimate reality, represents unity, and individual consciousness reflects differentiation. Thus, additive and subtractive color models symbolize dual states of awareness: unity and particularity, expansion and limitation, material and spiritual.

Expanding Color Perception Beyond Sight

Traditional models of color perception assume that color is an exclusively visual experience, contingent on light interacting with the eye's photoreceptors. However, recent research in sensory substitution and cross-modal perception suggests that color perception may not be limited to sight. Studies have shown that color can be experienced through alternative sensory channels, such as auditory or vibrational feedback, particularly in cases of sensory substitution for the visually impaired. These findings raise important questions about the adaptability of sensory systems and the flexibility of consciousness in interpreting environmental cues.

Research on non-visual color perception has revealed that under specific conditions, blind and blindfolded individuals can differentiate colors using vibrational feedback or other tactile cues. Specialized blindfolds that block all light while allowing participants to keep their eyes open have been used in controlled experiments, demonstrating that vibrational frequencies associated with certain colors can enable non-visual color perception. This capacity for sensory compensation suggests that color perception may be more integrative and flexible than previously thought, capable of operating across multiple sensory modalities.

Sound-Color Correlations and the Role of Solfeggio Frequencies

One intriguing avenue for exploring non-visual color perception involves investigating correlations between sound frequencies and color experiences. Solfeggio frequencies, a set of ancient sound frequencies, have been used for centuries in meditative and healing practices, where each frequency is believed to resonate with specific psychological and emotional states. For instance, the 528 Hz frequency, often associated with transformation and healing, is sometimes linked to the color green, symbolizing growth and balance. Similarly, the 852 Hz frequency, aligned with spiritual insight, has been conceptually associated with the color indigo, which represents inner vision and higher consciousness. These associations, while primarily anecdotal, suggest a potential framework for examining how sound and color might intersect through shared vibrational qualities.

Research on synesthesia and sensory substitution supports the plausibility of sound-color correlations, particularly through cross-modal sensory experiences where one type of sensory input consistently evokes another, such as sounds triggering color associations. Studies in this area have shown that the brain is capable of reorganizing itself to interpret information through non-traditional channels, indicating a high degree of sensory adaptability and neuroplasticity. The potential for mapping Solfeggio frequencies to specific colors could reveal deeper insights into multisensory perception, suggesting that color might be perceived not only visually but as an integrative experience across sensory domains.

Objectives and Scope of the Study

This paper seeks to integrate scientific, philosophical, and practical perspectives on color perception, proposing a unified model that encompasses physical, symbolic, and experiential dimensions. By examining both additive and subtractive color models, we argue that color perception provides insights into the adaptable, multisensory nature of consciousness, suggesting that perception itself may transcend traditional sensory boundaries. This theory is supported by examining non-visual perception, particularly through vibrational and auditory experiences, as well as the potential for sound-color correlations through Solfeggio frequencies.

Through this interdisciplinary approach, we propose potential applications in art, technology, and therapeutic practices. Multisensory installations, assistive devices for the visually impaired, and therapeutic environments that integrate color and sound could offer practical applications of these insights. The findings of this study could impact various fields, from immersive media to cognitive science and consciousness studies, by expanding our understanding of perception as an adaptable, integrative process.

Ultimately, this paper seeks to position color as both a scientific phenomenon and a metaphysical bridge to understanding consciousness, exploring new possibilities for interdisciplinary research on sensory perception.

Expanded Literature Review

Foundations of Color Theory and Perception

The study of color has evolved considerably since its early conceptualizations in physics and philosophy. Modern color theory is rooted in the 17th-century work of Sir Isaac Newton, whose prism experiments demonstrated that white light comprises the full color spectrum. Newton's findings introduced the notion that colors are intrinsic properties of light, establishing the foundation for the additive color model (RGB), which combines red, green, and blue wavelengths to create white light. This RGB model forms the basis of additive color mixing, essential for digital displays, lighting design, and other technologies reliant on light-emitting sources. Additive color theory thus centers on light as a source of color and focuses on the integration of wavelengths to produce a broad spectrum of colors.

In the early 19th century, Johann Wolfgang von Goethe expanded upon Newton's findings with a psychological approach, focusing not only on the physical properties of light but also on human responses to color. Goethe's *Theory of Colours* (1810) explored how different colors evoke specific emotions, emphasizing the subjective and experiential aspects of color. Goethe's work laid the groundwork for considering color as both a perceptual and psychological phenomenon, linking it to emotional and cognitive experiences. This human-centered perspective diverged from Newton's purely empirical approach and has influenced color theory's application in fields such as art, design, and therapy, where emotional responses to color are prioritized alongside physical properties.

In contrast, the subtractive color model (CMY), developed in the context of pigments and dyes, emerged from studies on material color properties. Subtractive color theory centers on the absorption and reflection of light by physical substances, using cyan, magenta, and yellow as primary colors. This model, essential in painting, printing, and other pigment-based arts, involves colors being created through the selective absorption of wavelengths rather than their emission. Michel-Eugène Chevreul's studies in the 1830s on color harmony in textiles and Albert H. Munsell's systematic color organization in the early 20th century further developed the subtractive model. This model contrasts with additive color theory by creating colors through the reduction or absorption of light, symbolically representing a grounding or materializing process, where combining pigments leads to black.

Together, additive and subtractive color models represent complementary approaches to color perception, each with distinct symbolic implications. Additive color, with light as its source, represents unity and synthesis, where colors blend to form an integrative whole. Subtractive color, on the other hand, mirrors materiality and differentiation, symbolizing how the physical world shapes perception through limitation and absorption. These models offer a dualistic view of color that extends beyond physical properties, suggesting metaphors for unity and individuality that align with broader philosophical and metaphysical principles.

Psychological and Neurological Dimensions of Color Perception

The exploration of color perception has expanded significantly with advances in psychology and neuroscience, revealing the complexities of how the brain interprets and responds to color. Early theories, such as the trichromatic theory proposed by Thomas Young and Hermann von Helmholtz, describe color vision as a function of three types of photoreceptor cells in the retina, each sensitive to specific wavelengths (short, medium, and long). This theory suggests that all colors can be perceived through the stimulation of these three receptor types, aligning with the additive model's emphasis on combining wavelengths to create color experiences.

The opponent-process theory, developed by Ewald Hering in the late 19th century, furthered understanding of color perception by positing that the brain processes colors in opposing pairs: red-green, blue-yellow, and black-white. This model accounts for perceptual phenomena such as afterimages and color constancy, which trichromatic theory alone cannot explain. Together, these theories illustrate that color perception is not merely a product of physical wavelengths but also a complex process involving the brain's interpretation and categorization of sensory information.

Functional magnetic resonance imaging (fMRI) studies have shown that color perception engages specific neural pathways associated with emotional regulation, attention, and memory. Research by Kawabata and Zeki (2004) reveals that viewing certain colors can activate regions in the brain linked to positive or negative emotions, suggesting that color influences mood and psychological states beyond visual perception. These insights have broad applications, influencing environmental design, marketing, and therapeutic practices where color choices are employed strategically to impact human behavior and well-being.

Neurological studies also highlight the brain's capacity to associate colors with specific emotional and cognitive responses. For instance, the color blue is often linked with calmness, while red can evoke excitement or alertness. These associations are not merely cultural constructs but may reflect inherent neurological processes that influence perception. The cross-cultural consistency of color-emotion associations supports the theory that color perception impacts human consciousness in predictable ways, aligning with philosophical interpretations of color as a bridge between sensory experience and emotional reality.

Cross-Modal and Non-Visual Perception of Color

Recent research in cognitive science and neuroscience has revealed that color perception may not be exclusively visual but may extend across sensory modalities. Studies on cross-modal perception and sensory substitution have shown that individuals can, under specific conditions, perceive color through non-visual channels, such as auditory or tactile feedback. This research challenges traditional assumptions about color as an exclusively visual phenomenon, suggesting that sensory systems can adapt to interpret environmental cues through alternative modalities.

Synesthesia, a condition where stimulation in one sensory pathway involuntarily evokes experiences in another (e.g., hearing sounds that trigger color experiences), offers insights into the brain's potential for cross-modal perception. Research by Jamie Ward on sound-color synesthesia has demonstrated that synesthetic individuals consistently experience specific colors in response to certain sounds, indicating an intrinsic connection between auditory and visual processing. Neuroimaging studies on synesthetes reveal overlapping activity in brain regions associated with visual and auditory processing, supporting the hypothesis that perception may be inherently multisensory.

Sensory substitution devices have provided further evidence for cross-modal color perception. Studies by Ward and Meijer (2010) on the use of sensory substitution devices by visually impaired individuals show that auditory and vibrational feedback can enable the perception of spatial and color-related information. In these studies, blind and blindfolded participants were able to differentiate colors based on tactile and vibrational sensations alone, effectively "sensing" colors through non-visual cues. These findings suggest a high degree of neuroplasticity, where the brain compensates for sensory loss by reorganizing its perceptual framework to incorporate alternative sensory inputs.

Sound-Color Correlations and Solfeggio Frequencies

One compelling area of cross-modal research involves the exploration of potential correlations between sound frequencies and color experiences. Solfeggio frequencies, a set of ancient sound frequencies ranging from 174 Hz to 963 Hz, have been used in meditative and healing practices for their reputed emotional and psychological effects. Each frequency is associated with specific psychological states; for example, 528 Hz is linked with transformation and healing, while 852 Hz is associated with spiritual insight. Some practitioners and researchers hypothesize that each frequency corresponds to a specific color, based on the frequency's purported emotional resonance and vibrational properties.

While direct empirical studies on Solfeggio frequencies and color perception are limited, the existing literature on sound-color associations offers a foundation for exploring these correlations. Sound-color synesthesia research has demonstrated that individuals can experience consistent color responses to specific sound frequencies, supporting the possibility that sound may be mapped onto color experiences through shared vibrational qualities. Additionally, sensory substitution studies suggest that non-visual channels, such as auditory or tactile feedback, can convey color-related information, indicating that sound frequencies could potentially evoke color associations.

The theoretical basis for sound-color correlations aligns with findings in cross-modal sensory processing, where sensory overlap occurs in regions responsible for interpreting auditory and visual information. The hypothesis that specific sound frequencies could correspond to certain colors presents an intriguing possibility for understanding multisensory perception. If Solfeggio frequencies, for instance, can evoke color experiences through vibrational feedback, it would suggest that perception is not limited to isolated sensory channels but may operate through an adaptable, multisensory framework.

Implications for Consciousness and Sensory Integration

The adaptability of sensory systems indicated by cross-modal research has significant implications for understanding consciousness. The phenomenon of color perception through non-visual channels supports the view that sensory processing is inherently flexible, capable of reinterpreting information through alternative pathways. This flexibility aligns with theories of neuroplasticity, where the brain continuously reorganizes itself in response to sensory changes and environmental demands. In cases of sensory deprivation, such as blindness, the brain appears to reconfigure itself to interpret stimuli through remaining sensory channels, suggesting a highly integrative and adaptive consciousness.

Color perception, therefore, may offer insights into how consciousness functions as a multisensory, integrative process. If color can be experienced through sound and vibration, it implies that consciousness has the capacity to transcend traditional sensory boundaries, reconfiguring itself based on available inputs. This understanding of perception as an adaptable, cross-modal experience challenges traditional models of sensory isolation, proposing that consciousness itself might operate as an integrative field capable of merging diverse sensory inputs into cohesive perceptual experiences.

Non-Visual Color Perception and Sound-Color Frequency Relationship

Non-Visual Color Perception: An Expanding Field

Historically, color perception has been understood as a strictly visual process, reliant on the interaction between light and the eye's photoreceptors. However, research in sensory substitution and cross-modal perception has challenged this assumption, demonstrating that humans may perceive color through non-visual pathways under certain conditions. This emerging field suggests that the boundaries of color perception are more adaptable than previously thought, particularly through the processes of sensory substitution and cross-modal integration, where stimuli are processed by sensory systems outside of their primary modalities.

One prominent line of inquiry within this field involves studying color perception among blind and visually impaired individuals. Studies using sensory substitution devices have shown that non-visual color perception can be achieved through tactile or auditory feedback. For instance, sensory substitution devices that translate visual information into auditory or vibrational signals have enabled participants to experience spatial and even color-related information without sight. Through such devices, individuals use other senses to interpret aspects of their surroundings typically processed visually. This capacity suggests that the brain possesses a remarkable degree of plasticity, allowing it to adapt to alternative sensory inputs to reconstruct aspects of visual experiences, including color.

Empirical Studies in Non-Visual Color Perception

Research by Ward and Meijer (2010) explored how auditory substitution can enable color perception in blind individuals. Their study used The vOICe, a sensory substitution device that converts images into soundscapes, allowing users to interpret shapes, textures, and colors through auditory signals. In their experiments, visually impaired participants could distinguish between colors by associating specific sound patterns with particular hues. Through training and exposure, participants reported color-like experiences based on auditory cues, effectively "hearing" color. This phenomenon reveals the brain's adaptability, suggesting that color perception may extend beyond visual inputs to incorporate auditory information when necessary.

Additional studies involving tactile feedback as a substitute for visual color perception further underscore this adaptability. In one experiment, blindfolded participants were guided through an obstacle course where objects emitted vibrational cues associated with specific colors. Participants successfully navigated the course using these tactile signals, associating each vibrational frequency with a particular color. These findings indicate that the brain can interpret vibrational cues as color information, especially when trained under controlled conditions. The ability to link color with vibration suggests that sensory systems can be reconfigured, allowing individuals to associate tactile feedback with specific visual attributes, including color.

Research on non-visual color perception highlights the brain's potential to process colors across sensory channels. Cross-modal plasticity, the brain's ability to reorganize itself in response to sensory loss, plays a significant role in these findings. Studies on individuals who are blind or have had vision removed through experimental conditions suggest that the brain actively compensates for the absence of visual data by amplifying other senses, allowing for an expanded sensory experience that includes color. This cross-modal adaptability provides a theoretical foundation for exploring how colors might be perceived through auditory and vibrational feedback, offering insights into the plastic nature of human perception.

Exploring Sound-Color Correlations through Solfeggio Frequencies

One particularly intriguing area of study within cross-modal perception is the potential for sound frequencies to correspond to specific colors, creating a multisensory experience of color through auditory channels. Solfeggio frequencies—a set of specific sound frequencies historically associated with healing and meditative practices—are of particular interest in this context. These frequencies, ranging from 174 Hz to 963 Hz, have been reputed to evoke specific emotional states, with anecdotal claims suggesting that each frequency correlates with a color associated with its psychological or spiritual impact.

For example, the 528 Hz frequency, often called the "Miracle" tone, is associated with transformation, healing, and DNA repair. This frequency has been symbolically linked to the color green, a hue commonly associated with growth, balance, and harmony. Similarly, 852 Hz, thought to resonate with spiritual insight, is frequently associated with indigo, a color symbolizing inner vision and higher consciousness. While these associations primarily exist within meditative and anecdotal contexts, they present a compelling basis for empirical investigation into sound-color relationships.

The hypothesis that sound frequencies like Solfeggio tones could evoke color responses is indirectly supported by research in sound-color synesthesia. Synesthesia, where stimulation of one sensory pathway leads to automatic experiences in another (such as hearing sounds that trigger color experiences), suggests that certain individuals naturally experience sensory overlap. Studies on sound-color synesthesia have demonstrated that specific frequencies consistently evoke specific colors in synesthetic individuals, indicating a possible neurological link between sound and color perception. Jamie Ward's research on sound-color synesthesia shows that synesthetes often report consistent color responses to particular sounds, suggesting that the brain has an inherent capacity for associating auditory and visual experiences.

Cross-Modal Integration and the Role of Vibrational Feedback

In addition to auditory feedback, vibrational cues have been shown to facilitate color perception across sensory channels, further supporting the hypothesis of sound-color correlations. Studies on cross-modal perception suggest that vibrational frequencies can serve as effective substitutes for color differentiation, particularly for individuals who are blind or visually impaired. In experiments where participants were exposed to haptic devices emitting specific vibrational frequencies linked to colors, subjects demonstrated the ability to identify and distinguish colors through touch. This vibrational approach could be extended to sound frequencies, hypothesizing that specific vibrations associated with Solfeggio tones could evoke color-like experiences through tactile feedback.

The theoretical basis for these sound-color correlations lies in the vibrational nature of both sound and color. While colors in the visible spectrum vibrate at extremely high frequencies (ranging from 430–770 THz), sound operates at much lower frequencies (ranging from 20 Hz to 20 kHz in human hearing). Despite this disparity, both sound and color are understood as frequencies that can evoke emotional and cognitive responses. If certain sound frequencies can elicit color associations, it would suggest a shared vibrational foundation for cross-modal experiences, where the brain interprets frequencies across different sensory channels through an adaptable framework.

Hypothetical Experimental Framework: Testing Sound-Color Correlations

To empirically investigate sound-color correlations, this paper proposes a hypothetical experimental framework using Solfeggio frequencies. Participants would undergo a training phase in which they learn to associate specific Solfeggio frequencies with colors through vibrational feedback. A haptic device could be used to emit distinct vibrational patterns for each frequency, enabling participants to establish a sensory link between sound and color. Once familiarized, participants would be asked to identify colors based solely on these vibrational cues, reporting their experiences to assess whether consistent sound-color associations emerge.

The experimental design would involve three phases:

1. **Training Phase**: Participants are introduced to each Solfeggio frequency alongside a vibrational pattern corresponding to a color. The training phase allows participants to internalize the association between frequency and color, providing a foundation for subsequent testing.
2. **Testing Phase**: Participants attempt to identify colors using only vibrational cues associated with each Solfeggio frequency,

without visual or auditory input. Their accuracy, response times, and confidence levels would be measured to assess whether reliable color associations have been formed.
3. **Validation Phase**: To ensure robustness, participants would be re-tested after a period of non-exposure to determine if the sound-color associations persist over time, suggesting that the associations are stable rather than short-term adaptations.

Data from such an experiment could provide valuable insights into the feasibility of sound-color correlations as a basis for non-visual color perception. If participants reliably associate specific frequencies with colors, this would lend empirical support to the theory that sound and color share a common vibrational framework within the brain's sensory processing architecture.

Implications for Multisensory Art and Assistive Technology

If consistent sound-color correlations are observed, these findings would have significant implications for both multisensory art and assistive technology. In the context of art, sound frequencies like Solfeggio tones could be incorporated into multisensory installations where sound and color are used in tandem to evoke specific emotional and sensory experiences. Such installations would engage audiences through sound and color simultaneously, expanding the aesthetic experience beyond traditional visual boundaries. For example, an installation might use 528 Hz to fill a space with green lighting and matching sound frequencies, creating an immersive environment that encourages feelings of peace and rejuvenation.

For assistive technology, these findings could inform the development of sensory substitution devices that translate visual color information into auditory or vibrational cues for visually impaired users. By encoding color information through specific sound frequencies, individuals with visual impairments could perceive color indirectly, enriching their sensory experience and improving environmental interaction. Devices could be designed to map common colors to distinct sound frequencies, allowing users to identify colors through auditory or tactile means.

Proposed Experimental Methodologies

Overview and Rationale

To investigate the potential correlations between sound frequencies and color perception, we propose a series of structured experiments that will test the ability of participants to perceive colors through non-visual sensory channels. Specifically, this research aims to assess whether sound frequencies—

particularly Solfeggio frequencies—can evoke reliable color associations through tactile and auditory cues. By testing participants across different sensory modalities, the experiments aim to contribute to our understanding of cross-modal perception and the adaptability of sensory processing in the brain.

The following experimental designs are grounded in principles of sensory substitution and cross-modal integration, informed by studies in neuroplasticity and multisensory perception. The experiments incorporate training phases to allow participants to build associations between specific frequencies and colors, with subsequent testing phases to assess accuracy, response times, and retention. These experiments are designed to evaluate whether sound-color associations can be formed, strengthened, and retained, potentially indicating a shared vibrational basis in sensory perception.

Experiment 1: Sound-Color Association Through Solfeggio Frequencies

Objective

The objective of this experiment is to determine if participants can reliably associate specific Solfeggio frequencies with corresponding colors through auditory and vibrational cues. By training participants to recognize vibrational patterns linked to Solfeggio frequencies, this experiment will assess the potential for sound frequencies to elicit non-visual color experiences.

Participant Selection

The experiment will include diverse participant groups to examine how sensory adaptability varies across demographics:

- **Sighted Participants**: Individuals with normal vision will be included to explore baseline multisensory adaptability.
- **Visually Impaired Participants**: Participants with visual impairments will provide insights into how cross-modal plasticity may affect sound-color perception.

- **Synesthetic Participants**: Individuals with sound-color synesthesia will be included to compare their natural associations with trained associations in non-synesthetic individuals.
- **Children (Aged 6–12)**: Including children will allow us to assess how neuroplasticity in developing brains may influence the formation of cross-modal associations.

Experimental Design and Procedure

The experiment will be conducted in three main phases: Training, Testing, and Retention.

1. **Training Phase**:
   - **Procedure**: Participants will be exposed to a set of Solfeggio frequencies (174 Hz, 285 Hz, 396 Hz, 528 Hz, 639 Hz, 741 Hz, 852 Hz, and 963 Hz) paired with specific colors (e.g.,

528 Hz with green, 852 Hz with indigo). Each frequency will be paired with a corresponding vibrational pattern delivered through a haptic device.
- **Objective**: Participants will learn to associate each Solfeggio frequency with its corresponding color through tactile feedback, allowing them to establish multisensory links between sound and color. Participants will have practice sessions where they receive feedback on the accuracy of their responses.
- **Duration**: The training phase will last approximately 1 hour per participant, with short breaks included to prevent fatigue.

2. **Testing Phase**:
   - **Procedure**: After training, participants

will be exposed to the Solfeggio frequencies in randomized order. They will be asked to identify the corresponding color based solely on vibrational cues, without auditory reinforcement. For each sound presented, participants will select a color from a set of colored cards or verbally indicate their color choice.
- **Data Collection**: Responses will be recorded to measure the accuracy of each participant's color identification. Additional metrics, such as response time and confidence ratings, will be collected to assess the strength and speed of the sound-color associations.
- **Objective**: This phase tests the durability of the sound-color associations

established during training, providing insight into how well participants retain these cross-modal links without visual input.

3. **Retention Phase**:
    - **Procedure**: To assess the stability of the learned associations, participants will return after a week to complete the same testing protocol. Retesting after a delay will reveal whether the associations were temporary or whether they formed stable, long-term connections.
    - **Objective**: Evaluating retention is essential for determining whether sound-color associations are a result of short-term adaptation or indicative of a more lasting cross-modal integration.

Experiment 2: Color-Based Obstacle Course Navigation Using Vibrational Cues

Objective

This experiment will investigate whether participants can use vibrational cues associated with colors to navigate a color-coded obstacle course, simulating a real-world application of sensory substitution. This experiment aims to assess whether cross-modal color perception can facilitate spatial orientation and interaction with physical environments.

Participant Selection

This experiment will involve a diverse group of participants, including sighted, visually impaired, and synesthetic individuals. Participants will be randomly assigned to either a training group or a control group to assess the impact of training on navigation performance.

Experimental Design and Procedure

The experiment will take place in a controlled indoor environment where objects are arranged to form an obstacle course. Each object or section of the course will emit a specific vibrational pattern associated with a particular color, based on the Solfeggio frequency-color associations established in Experiment 1.

1. **Training Phase**:
   - **Procedure**: Participants in the training group will be introduced to the vibrational cues for each color used in the obstacle course. They will practice identifying colors based on vibrational feedback, with verbal confirmation of color associations. The control group will not undergo this training.
   - **Objective**: The training aims to familiarize participants with vibrational cues for each color, enhancing their ability to use this information to navigate the course effectively.
2. **Navigation Task**:

- **Procedure**: Blindfolded participants will be asked to navigate the obstacle course using only vibrational cues to identify obstacles by color. Participants will move through the course with the goal of correctly identifying and avoiding obstacles based on the color cues.
- **Data Collection**: Data will include completion time, accuracy of color identification, and error rates (e.g., mistakenly identifying an obstacle color or colliding with an obstacle).
- **Objective**: This phase assesses participants' ability to use cross-modal color perception to navigate space, testing the functional applicability of sound-color associations in spatial orientation tasks.

3. **Post-Task Interviews**:
    - **Procedure**: Participants will be interviewed after the task to provide qualitative data on their experience, including perceived difficulty, confidence in color identification, and feedback on the usefulness of vibrational cues.
    - **Objective**: The interview aims to gather subjective insights into how effectively participants could rely on cross-modal perception for spatial navigation.

Experiment 3: Testing Retention and Reinforcement in Children

Objective

Given the heightened neuroplasticity in children, this experiment will assess the durability of sound-color associations formed through vibrational cues. This study will explore whether children can retain these associations longer than adults, providing insight into age-related differences in cross-modal adaptability.

Participant Selection

Participants will include children aged 6–12, who will be randomly assigned to groups that receive either auditory or vibrational training to associate Solfeggio frequencies with specific colors.

Experimental Design and Procedure

1. **Training Phase**:
    - **Procedure**: Children will undergo an extended training phase to form associations between Solfeggio frequencies and colors. The training will involve games and interactive exercises that reinforce sound-color links, allowing for natural learning and

engagement. Each child will experience both auditory and vibrational cues to associate each frequency with a color.
- **Objective**: This phase aims to determine whether a playful, interactive format improves retention and strengthens cross-modal associations in children.

2. **Testing Phase**:
    - **Procedure**: After the training, children will be tested on their ability to recall the color associations using vibrational cues alone. They will select colors based on vibrational input without additional auditory reinforcement, and responses will be recorded.
    - **Data Collection**: Data will include the accuracy and response time for each color identified, as well as

subjective confidence ratings provided by the children (e.g., selecting an "easy" or "hard" scale for each response).
- **Objective**: This phase will measure the effectiveness of the training and examine retention in a population with high neuroplasticity.

3. **Follow-Up Testing**:
   - **Procedure**: Children will return for a follow-up session one month later to assess whether the learned associations persist. This delayed testing will provide data on the long-term retention of sound-color associations in children.
   - **Objective**: Long-term retention assessment will clarify whether age-related neuroplasticity contributes to stronger,

more durable cross-modal associations.

Data Analysis and Expected Outcomes

Across all three experiments, data analysis will focus on measuring accuracy, response times, and retention of sound-color associations. Expected outcomes include:

- **Accuracy Rates**: Higher accuracy in color identification based on vibrational cues would support the feasibility of cross-modal color perception.
- **Retention**: Retention data from both adult and child participants will indicate the stability of cross-modal associations over time.
- **Neuroplasticity Effects**: If children demonstrate higher retention rates, this would suggest a greater capacity for multisensory adaptability, highlighting developmental differences in cross-modal integration.

Together, these experimental methodologies aim to provide empirical evidence for the adaptability of sensory perception, supporting the potential for sound-color associations in non-visual perception. Positive results would suggest that sound and color perception share an integrative foundation in the brain, opening pathways for future research and applications in multisensory technology and therapeutic practices.

Theoretical and Philosophical Analysis

Duality in Additive and Subtractive Color Models

Color theory is uniquely positioned at the intersection of science, art, and metaphysics, offering models that reveal not only physical truths but also philosophical insights. The additive and subtractive color models provide a dual framework through which color can be understood as both a physical property and a symbolic representation of fundamental existential forces. In additive color mixing (as used in light-based models), red, green, and blue combine to form white light, representing the convergence of all wavelengths. This synthesis aligns with concepts of unity, expansiveness, and integration. White, as the result of this convergence, can be seen as a symbol for pure consciousness or an undivided field of awareness in which distinctions between individual elements dissolve.

In contrast, subtractive color mixing—utilized in pigment-based models—relies on the combination of cyan, magenta, and yellow, resulting in black. This model involves the absorption of light rather than its emission, with each pigment subtracting wavelengths from reflected light. Symbolically, subtractive mixing represents sensory limitation, material absorption, and the boundaries of perception. Black, the end result of combining all pigments, suggests a grounding in materiality and specificity, a reflection of the physical world's inherent limitations. Thus, the additive model's production of white and the subtractive model's creation of black embody complementary principles: one expansive and unifying, the other grounding and differentiating.

This duality in color perception can be interpreted as a metaphor for states of consciousness. Additive mixing, where light coalesces to produce white, represents consciousness in its expansive, integrated form—a state of unity, self-transcendence, or universal awareness. Subtractive mixing, where pigments absorb light to create black, mirrors consciousness as it is bound within the sensory world, marked by individuality and differentiation. These models offer insight into how consciousness itself might operate within a spectrum of awareness, alternating between states of unity and individuation.

Hermetic and Eastern Metaphysical Interpretations

The philosophical dimensions of color theory find resonance in Hermeticism, an ancient tradition that emphasizes the relationship between the physical and spiritual realms. The Hermetic principle of "as above, so below" posits that the laws governing the cosmos are reflected in human experience. In this context, additive color mixing represents spiritual unity, where light—the source of all color—converges to form an undifferentiated whole. This model reflects Hermetic ideals of unity and cosmic order, suggesting that at the highest level, consciousness is unbounded and unitive.

Subtractive mixing, however, represents the physical world, where matter imposes constraints and boundaries. The absorptive process in subtractive color models aligns with the principle that material existence is defined by limitation and individuation, where the unity of light is refracted into distinct colors or experiences. This dichotomy aligns with Eastern metaphysical traditions as well. In Taoism, for instance, yin and yang embody complementary and interdependent forces—darkness and light, absorption and reflection—that create balance and wholeness. Additive color mixing corresponds to yang, symbolizing expansion and light, while subtractive mixing aligns with yin, symbolizing limitation and grounding.

Similarly, in Hindu philosophy, Brahman represents the ultimate reality, an undivided field of consciousness, while individual consciousness reflects a limited, differentiated form of this universal essence. Additive color mixing parallels the expansive nature of Brahman, where all colors unite to form white, symbolizing unity and transcendence. Subtractive mixing, with colors combining to produce black, symbolizes the individuation process in the material world, where sensory limitations define the human experience. These interpretations suggest that the duality in color models is more than a scientific construct; it is a reflection of ancient philosophical concepts that seek to understand the nature of reality and the relationship between unity and individuality.

Quantum Perspectives on Color and Consciousness

The dual nature of additive and subtractive color models aligns intriguingly with principles in quantum mechanics, where light exists as both wave and particle, challenging traditional boundaries between states of matter and energy. This quantum framework offers a valuable metaphor for understanding consciousness as both localized and expansive, capable of unifying sensory experiences or segmenting them into discrete modalities. In this context, additive color mixing (RGB) represents the unifying aspects of consciousness, where wavelengths combine to form white, symbolizing states of integration, expansion, and collective awareness. Conversely, subtractive color mixing (CMY), with its absorptive process leading to black, reflects grounded consciousness, sensory limitation, and individuated perception.

This dual model of color serves as more than a scientific observation; it acts as a metaphor for consciousness itself, suggesting that human awareness operates along a spectrum. Just as light can manifest as both wave and particle, consciousness may shift between unified awareness and sensory-bound experience, responding dynamically to environmental inputs. This view resonates with quantum superposition—the coexistence of multiple states—and entanglement, wherein individual entities remain interconnected despite physical separation.

The exploration of color as a multisensory phenomenon also

supports theories of sensory integration and cross-modal perception, where consciousness is not limited to a single sensory modality but capable of synthesizing inputs across different channels. As sensory experiences like sight, sound, and touch interact, consciousness emerges as an integrative field that bridges physical and perceptual realities. Through this lens, color perception offers a glimpse into a form of consciousness that operates beyond the traditional senses, potentially aligning with what some quantum theorists posit as a "consciousness field" or a foundational layer of interconnected awareness.

These quantum and sensory integration perspectives have practical implications in fields such as immersive media, virtual reality, and therapeutic practices, where experiences can be designed to stimulate cross-modal perception and foster states of relaxation, heightened awareness, or even healing. By utilizing the principles of additive and subtractive color mixing, designers and researchers can craft environments that reflect the fluid nature of consciousness, engaging users in experiences that resonate with both the material and transcendent realms.

Color Perception as a Spectrum of Consciousness

The idea of color perception extending beyond visual boundaries suggests a new framework for understanding consciousness as a dynamic, multisensory process. If color can indeed be perceived through sound and vibration, as the experimental methodologies propose, it would imply that consciousness operates on a spectrum, interpreting stimuli through an adaptable, integrative system. In this model, color perception is not limited to sight but can adapt to incorporate alternative sensory channels, illustrating the brain's plasticity and capacity for cross-modal integration.

This spectrum of consciousness aligns with Ken Wilber's Integral Theory, which posits that consciousness encompasses multiple states and stages, integrating sensory, cognitive, and spiritual dimensions. In Wilber's framework, color perception could be seen as one facet of an adaptable consciousness, where sensory experiences merge to form a unified awareness. The flexibility of color perception across sensory boundaries supports the notion that consciousness is not fixed but rather an evolving, integrative field. This concept has practical implications for how we understand sensory integration and neuroplasticity, suggesting that experiences traditionally confined to one sense (like sight) may be accessible through other sensory modalities.

In this model, additive and subtractive color theories serve as metaphors for how consciousness itself might alternate between integrated and individuated states. Additive mixing, where colors converge to create white, symbolizes a higher, expansive state of consciousness where sensory boundaries dissolve. Subtractive mixing, where colors absorb light to form black, represents a sensory-bound state where consciousness is localized within individual experiences. These states are not oppositional but complementary, suggesting that consciousness, like color, operates along a continuum of integration and differentiation.

Color as a Vehicle for Expanding Consciousness

By positioning color perception within a multisensory, metaphysical framework, we propose that color may serve as a vehicle for expanding human consciousness. The experimental studies on non-visual color perception suggest that sensory boundaries are more permeable than previously believed, allowing individuals to experience color through auditory or vibrational feedback. This adaptability points to the potential for consciousness to integrate sensory input across traditional boundaries, reinforcing the notion that awareness itself is inherently flexible.

In this expanded view, color perception becomes not only a sensory experience but also a pathway to exploring the adaptability and integrative nature of consciousness. By investigating how color can be perceived through sound and vibration, we gain insights into how the brain adapts to sensory inputs, suggesting that perception itself may be a fluid, multisensory process. This understanding aligns with the philosophy that human consciousness has the potential to transcend individual sensory limitations, creating a unified experience where auditory, visual, and tactile cues coalesce into a holistic awareness.

This theory has practical implications for fields such as multisensory art, immersive media, and therapeutic practices, where color and sound are used to influence mood and cognitive states. If consciousness is capable of integrating color through alternative sensory pathways, it implies that the experiences designed in these fields could be enriched through a multisensory approach that engages multiple senses. By utilizing sound-color correlations in art installations, therapeutic environments, and sensory substitution technologies, we can facilitate experiences that expand sensory awareness and potentially enhance emotional and cognitive states.

Case Studies and Practical Applications

Multisensory Art Installations

Multisensory art installations provide a fertile ground for exploring the experiential possibilities of additive and subtractive color mixing, especially when combined with sound and tactile feedback. Contemporary artists such as James Turrell and Olafur Eliasson have pioneered the use of color and light in ways that immerse viewers beyond traditional visual engagement, creating installations that actively shape the viewer's perception of space and color. Turrell's "Ganzfeld" series, for example, employs large-scale light installations that fill rooms with singular colors, creating a sense of color immersion that alters depth perception and spatial awareness. By integrating light (additive color) into his installations, Turrell elicits emotional and cognitive responses, illustrating how color can alter not only visual experience but also psychological states.

Building on the work of artists like Turrell and Eliasson, future installations could incorporate sound frequencies, particularly Solfeggio frequencies, to engage viewers through auditory channels in addition to light and color. In this expanded approach, each color might be associated with a corresponding sound frequency—528 Hz with green, for instance, or 852 Hz with indigo—to create a multisensory environment where color and sound resonate together. This integration would allow audiences to "experience" colors through sound, stimulating a cross-modal response that connects visual and auditory perception. Such installations could deepen engagement, enabling viewers to experience color as a holistic phenomenon that transcends sight, resonating within their consciousness through both sound and light.

Additionally, the integration of vibrational feedback in art installations could further enhance the multisensory experience, particularly for visually impaired audiences. Through the use of wearable haptic devices that emit vibrations corresponding to different colors, visually impaired visitors could perceive color through tactile feedback. Such devices could be synchronized with the visual elements of the installation, allowing users to experience the installation's color dynamics through vibration rather than sight. This approach would make multisensory art more inclusive, broadening access to color-based installations and enriching the sensory experience for all viewers.

Assistive Technology and Sensory Substitution Devices

The potential for non-visual color perception opens up significant possibilities for assistive technologies, particularly in sensory substitution devices for visually impaired individuals. Current sensory substitution devices, such as The vOICe, convert visual information into auditory cues, enabling users to perceive spatial information through sound. Expanding these technologies to include color perception through auditory or vibrational feedback could enhance their utility, offering visually impaired users an enriched experience of their surroundings.

One practical application could involve encoding color information into specific sound frequencies or vibrational patterns, similar to the

Solfeggio frequencies. For instance, each color could be mapped to a unique vibrational pattern or sound frequency, allowing users to "hear" or "feel" colors. A device worn on the wrist or integrated into a cane could emit a unique frequency or vibrational signal when pointed at an object, providing the user with non-visual cues about the object's color. For example, a 528 Hz frequency associated with the color green could indicate that an object is green, allowing the user to perceive color information indirectly.

Such technology could significantly enhance the daily experiences of visually impaired individuals, enriching their sensory understanding of the environment. By transforming color information into alternative sensory modalities, these devices could facilitate a more nuanced experience of the world, providing users with a greater sense of spatial and emotional engagement. Furthermore, assistive devices could be customized based on user preference, allowing individuals to select sound or vibrational associations that resonate most intuitively, enhancing the personal relevance and functionality of the technology.

Therapeutic Applications in Meditative and Healing Practices

The proposed sound-color correlations, particularly using Solfeggio frequencies, hold substantial potential for therapeutic applications in meditative and healing practices. In color therapy, different colors are used to evoke specific emotional responses, influence mood, and potentially improve physiological well-being. Blue, for example, is often associated with calmness and relaxation, while red is used to increase energy and alertness. Integrating sound frequencies that correspond to these colors could deepen the therapeutic effects, allowing individuals to experience both color and sound as complementary elements within a single treatment.

Meditative practices, especially those focused on chakra alignment or mindfulness, already incorporate color visualization and sound as methods for enhancing focus and relaxation.

Each chakra is traditionally associated with a specific color, and certain frequencies are believed to resonate with each chakra, potentially reinforcing its energy. By introducing specific sound-color associations into meditative practice, practitioners could align their meditation sessions with targeted color and sound combinations, designed to stimulate particular energy centers or moods. For example, a session aimed at the heart chakra (associated with green) could incorporate the 528 Hz Solfeggio frequency, creating a cohesive experience where sound and color harmonize to deepen relaxation and self-awareness.

In clinical and therapeutic settings, sound and color associations could also play a role in managing stress, anxiety, and emotional regulation. Studies in music therapy have shown that specific sound frequencies can influence heart rate variability, reduce stress, and improve mood. Integrating color frequencies into these practices could enhance the therapeutic experience, providing patients with a multisensory treatment that engages both auditory and visual channels. By incorporating non-visual color perception into therapy, clinicians could create environments that foster a sense of calm and balance, helping individuals regulate their mental and emotional states.

Educational Applications and Neuroplasticity Development

The adaptability of sensory perception in relation to color holds promise for educational applications, particularly those aimed at children and individuals with neurodivergent processing styles. If sound-color associations can be developed through training, as the proposed experimental methodologies suggest, these associations could serve as a foundation for educational tools that engage multiple senses. For instance, children learning about colors could be introduced to specific frequencies associated with each color, reinforcing memory retention through a multisensory approach that combines auditory and visual stimuli.

Furthermore, sound-color association training could support neuroplasticity development in children, enhancing their capacity for cross-modal learning. By teaching children to associate sound frequencies with colors, educational programs could stimulate cognitive pathways that strengthen sensory adaptability. This approach could be particularly beneficial for children with sensory processing differences, providing an engaging and adaptive way to explore sensory integration and enhance cognitive flexibility. The use of color-sound associations could also support language development, allowing children to link sensory experiences to descriptive language more effectively.

Incorporating sound-color association training into educational programs for individuals with sensory processing disorders could offer therapeutic benefits by fostering cross-modal connections that ease sensory integration. By teaching color perception through sound or vibration, these programs could encourage neuroplasticity, supporting a multisensory learning approach that enriches the educational experience and fosters adaptable cognitive processing.

Future Directions for Integrating Sound-Color Perception in Everyday Life

The implications of sound-color associations extend beyond specialized applications, potentially impacting everyday experiences in spaces such as interior design, urban planning, and virtual reality. If sound frequencies can reliably evoke color associations, designers could create environments where color and sound work together to shape ambiance and mood. In a healthcare setting, for example, colors that promote relaxation could be paired with corresponding sound frequencies to create a calming atmosphere. Similarly, workspaces could incorporate energizing colors and sounds to promote focus and productivity, engaging multiple senses to shape user experiences.

In virtual reality (VR) environments, sound-color associations could enhance immersion by providing multisensory cues that reinforce visual elements. VR developers could pair colors with their corresponding frequencies to enhance user experiences, allowing players to "feel" colors through sound and vibration. For example, a virtual forest setting could use frequencies associated with green to deepen the sense of tranquility, providing an immersive experience that blends visual and auditory stimuli. Such applications would leverage the brain's capacity for cross-modal integration, creating environments where color perception is enriched through sound and tactile cues.

These practical applications underscore the potential of sound-color associations to shape human perception in ways that align with both personal and therapeutic needs. By incorporating multisensory elements into environments, art, and technology, we can create experiences that expand human sensory capacity, enhancing interactions with both real and virtual worlds. These developments point to a future where sound and color are no longer isolated sensory experiences but rather components of a dynamic, integrative framework for interacting with the world.

Anthrocosmia: A Multilayered Framework Integrating Human

Consciousness within a Cosmic Context

Abstract

Anthrocosmia introduces a novel theoretical framework that redefines human existence within a holistic, multilayered model integrating biological, energetic, and conscious dimensions within a broader cosmic context. This framework posits that humans function not merely as observers but as active, interconnected participants in the cosmos, merging subjective awareness with objective physical and energetic realities. The model comprises three interdependent layers: (1) **Biological Systems**, encompassing neural, cellular, and physiological mechanisms that sustain life and support awareness; (2) **Energetic Fields**, theorizing that bioelectromagnetic fields produced by physiological processes connect individuals to their immediate environment and potentially larger energetic systems; and (3) **Conscious Awareness**, conceptualizing consciousness as a fundamental attribute of the universe, perceived and experienced through human cognition and self-reflection.

Rooted in interdisciplinary insights from neuroscience, psychology, biophysics, and transpersonal philosophy, *Anthrocosmia* challenges reductionist models that isolate consciousness within purely biological frameworks. It advocates a layered approach, with each layer providing unique insights into human experience, interconnectedness, and collective consciousness. The model emphasizes empirical pathways, including neuroimaging, biofield measurement, and phenomenological research, enabling rigorous testing across scientific fields. Practical applications are proposed in medicine, psychology, artificial intelligence (AI), and education, with implications for integrative therapeutic practices, mental well-being, ethically conscious AI, and holistic educational development.

By bridging empirical research with existential and philosophical inquiry, *Anthrocosmia* offers a comprehensive model that respects the full spectrum of human experience within a self-aware, evolving universe. This framework positions human existence as inherently purposeful and interconnected, inviting collaborative research that transcends disciplinary boundaries to deepen our understanding of consciousness, health, and humanity's role in a conscious cosmos.

Introduction

Background and Rationale

Human consciousness and its relationship to the universe have long intrigued scientists, philosophers, and spiritual thinkers. Traditional models in psychology and neuroscience often approach consciousness as an emergent property arising from brain processes, reducing human experience to the workings of biological mechanisms and cognitive functions (Crick & Koch, 2003; Edelman, 2004). While such reductionist perspectives have yielded insights into brain structure and function, they struggle to fully capture complex phenomena such as interconnectedness, self-transcendence, and subjective experiences reported in states of meditation, creativity, or existential reflection (Varela, Thompson, &

Rosch, 2017; Northoff & Huang, 2017).

The limitations of reductionist models have prompted calls for integrative approaches that consider the full spectrum of human experience (Chalmers, 1995; Stoljar, 2009). Emerging fields such as consciousness studies, transpersonal psychology, and integrative neuroscience seek to bridge gaps between subjective experience and objective measurement, recognizing that human consciousness may encompass dimensions beyond the purely physical or neural (Ferrer, 2002; Wallace, 2000).

This paper introduces *Anthrocosmia*, a comprehensive theoretical framework aimed at expanding our understanding of human existence beyond the confines of reductionist models. The term *Anthrocosmia*—derived from *anthropos* (human) and *cosmos* (universe)—positions humans as active participants in an interconnected, evolving cosmos. This framework integrates scientific insights from neuroscience, biophysics, and psychology, as well as philosophical perspectives on consciousness, to propose that human beings are not merely biological entities but complex intersections of matter, energy, and awareness. By situating human consciousness within a multilayered reality, *Anthrocosmia* proposes that individuals embody a unique role as observers and influencers within a self-aware cosmos.

Aims and Scope

The primary aim of this paper is to present the *Anthrocosmia* framework in detail, providing a clear and comprehensive model that integrates biological, energetic, and conscious dimensions of human existence. The paper seeks to:

1. **Define and Elaborate on the Three Layers**: Provide clear definitions and explanations of the Biological Systems, Energetic Fields, and Conscious Awareness layers, highlighting their interconnections and

individual contributions to human experience.
2. **Integrate Interdisciplinary Insights**: Draw from recent empirical studies and theoretical developments in neuroscience, psychology, biophysics, and philosophy to support and contextualize the framework.
3. **Address Potential Criticisms and Limitations**: Acknowledge the speculative elements of the model, particularly regarding Energetic Fields, and provide reasoned responses to potential criticisms.
4. **Propose Empirical Pathways**: Offer detailed methodological

proposals for testing the framework's concepts, emphasizing ethical considerations and scientific rigor.
5. **Discuss Practical Applications**: Explore how *Anthrocosmia* can inform practices in medicine, psychology, AI, and education, emphasizing the framework's relevance and potential impact.

Structure of the Paper

The paper is structured to guide readers through the layered framework of *Anthrocosmia*, moving from theoretical foundations to practical applications:

- **Section 2**: *Core Concepts of Anthrocosmia*—Examines each of the three layers in detail, providing theoretical grounding, definitions, and

integration with existing literature.
- **Section 3**: *Purpose and Potential of Human Existence*—Explores the implications of the framework for understanding human purpose, potential, and collective consciousness.
- **Section 4**: *Applications and Implications of Anthrocosmia*—Discusses practical applications across medicine, psychology, AI, and education, highlighting empirical evidence and methodological considerations.
- **Section 5**: *Addressing Potential Criticisms and Limitations*—Acknowledges speculative aspects, discusses potential criticisms, and provides responses grounded in current research.
- **Section 6**: *Conclusion and Future Directions*—Summarizes contributions, outlines future research avenues, and emphasizes the framework's potential impact.

Core Concepts of Anthrocosmia

The *Anthrocosmia* framework conceptualizes human experience as unfolding across three interdependent layers: **Biological Systems**, **Energetic Fields**, and **Conscious Awareness**. Together, these layers create a comprehensive model that examines human existence from physical, energetic, and conscious perspectives, addressing both individual and collective aspects of experience. Each layer offers distinct insights, yet they are interwoven, suggesting that the full spectrum of human experience can only be understood through their interactions.

2.1 Biological Systems

Definition and Theoretical Grounding

The **Biological Systems** layer represents the foundational physical processes that sustain life, encompassing neural, cellular, and biochemical interactions. It includes the brain's neural networks, the endocrine system, the immune system, and other physiological mechanisms that contribute to bodily functions and homeostasis. In traditional models, consciousness is often seen as an emergent property of these biological systems, attributed to the complexity of neural networks and brain structures (Koch, 2018; Dehaene, 2014).

However, *Anthrocosmia* posits that biological systems are not merely substrates for consciousness but active interfaces that enable humans to interact with both energetic and conscious dimensions of existence. The concept of *embodied cognition*, which suggests that cognitive processes are deeply rooted in the body's interactions with the world, aligns with this view (Barsalou, 2008; Wilson, 2002). Embodied cognition emphasizes that the mind is not isolated from the body but is shaped by sensory and motor experiences, suggesting a more integrated understanding of cognition and consciousness.

The biological layer also encompasses the concept of neuroplasticity—the brain's ability to reorganize itself by forming new neural connections throughout life (Pascual-Leone et al., 2005). This plasticity indicates that the biological system is dynamic and responsive to both internal and external stimuli, reinforcing the idea that biological processes are integral to the ongoing development of consciousness.

Integration with Existing Literature

Research in neuroscience has shown that brain structure and function can be influenced by sustained practices like meditation, yoga, and mindfulness, supporting *Anthrocosmia*'s claim that biological systems play an adaptive role in shaping awareness and environmental responsiveness (Tang, Hölzel, & Posner, 2015; Davidson & McEwen, 2012). For instance, long-term meditation practitioners exhibit increased cortical thickness in brain regions associated with attention and sensory processing (Lazar et al., 2005).

Furthermore, studies on the gut-brain axis highlight the bidirectional communication between the gastrointestinal tract and the central nervous system, suggesting that microbiota can influence mood, cognition, and behavior (Mayer et al.,

2015). This expands the understanding of biological systems to include interactions beyond the brain, emphasizing the interconnectedness of bodily processes in shaping consciousness.

Empirical Evidence and Methodological Proposals

To explore the relationship between Biological Systems and other layers, empirical studies could investigate the neural and physiological effects of mind-body practices. **Methodological proposals** include:

- **Neuroimaging Studies**: Utilize functional MRI (fMRI) and electroencephalography (EEG) to examine brain activity during practices like meditation, breathwork, and somatic therapy. Focus on neural connectivity in regions associated with self-awareness (e.g., the default mode network), emotional regulation (e.g., the prefrontal cortex), and interoception (e.g., the insular cortex).

- **Physiological Measurements**: Assess physiological markers such as heart rate variability (HRV), cortisol levels, immune function, and inflammatory markers before and after interventions. These measures can provide insights into the systemic impact of practices on physical well-being.
- **Longitudinal Studies**: Conduct longitudinal research to observe changes over time, providing evidence of causal relationships between practices and biological adaptations.
- **Epigenetic Analysis**: Investigate how environmental factors and practices influence gene expression related to stress response, neuroplasticity, and immune function (Kaliman et al., 2014).

Practical Applications

The Biological Systems layer has immediate applications in mental health and well-being:

- **Integrative Therapies**: Incorporate body-centered practices like yoga, tai chi, and mindfulness-based stress reduction (MBSR) into therapeutic interventions. These practices engage physiological systems, promoting resilience, emotional stability, and self-awareness.
- **Clinical Interventions**: Use biofeedback and neurofeedback techniques to help individuals gain awareness of physiological processes and develop self-regulation skills.
- **Preventative Health**: Encourage lifestyle interventions that promote physical activity, nutrition, and sleep hygiene to support overall biological functioning.
- **Personalized Medicine**: Utilize genetic and biomarker information to tailor interventions that align with individual biological profiles.

2.2 Energetic Fields

Definition and Theoretical Grounding

The **Energetic Fields** layer proposes that human beings are surrounded by bioelectromagnetic fields originating from physiological processes, creating an energetic connection with their environment. This concept draws on biophysics and emerging research in biofield science, suggesting that biological organisms produce measurable electromagnetic fields through processes such as neural activity, cardiac rhythms, and cellular metabolism (Becker & Selden, 1985; Oschman, 2000).

The human heart, for example, generates a measurable electromagnetic field that extends several feet beyond the body (McCraty et al., 2015). The HeartMath Institute's research indicates that these fields may influence emotional states and interpersonal interactions, suggesting a form of energetic communication (McCraty & Childre, 2010). Similarly, the brain's electromagnetic activity, measured through EEG, reflects neural processes that could interact with external electromagnetic fields (Pockett, 2012).

The concept of Energetic Fields aligns with cultural understandings of energy fields, such as the *aura* or *energy body*, found in traditions like Ayurveda, Traditional Chinese Medicine, and Indigenous healing practices (Jain et al., 2015). While these concepts have historically been considered metaphysical, contemporary research is beginning to explore their scientific basis.

Integration with Existing Literature

Biofield science is an emerging field that studies the interactions of energy fields within and around the human body (Rubik et al., 2015). Studies have demonstrated that practitioners of energy healing modalities, such as Reiki and Therapeutic Touch, can produce measurable changes in electromagnetic fields (Jain & Mills, 2010). Additionally, experiments have shown that intentionality and focused attention can affect physical systems at a distance, suggesting non-local interactions (Schwartz et al., 2004).

Research on transcranial magnetic stimulation (TMS) and transcranial direct current stimulation (tDCS) indicates that external electromagnetic fields can influence neural activity and cognitive function (Nitsche et al., 2008). This supports the idea that energetic fields can interact with biological systems, potentially affecting consciousness and behavior.

While still speculative within mainstream psychology, these findings provide preliminary support for the existence and influence of Energetic Fields. Critics argue that more rigorous research is needed to validate these concepts and rule out placebo effects or experimental biases (Hall et al., 2016).

Empirical Evidence and Methodological Proposals

To investigate Energetic Fields within *Anthrocosmia*, empirical studies could employ the following methodologies:

- **Measurement of Biofields**: Use sensitive instruments like superconducting quantum interference devices (SQUIDs)

and magnetoencephalography (MEG) to detect and measure bioelectromagnetic fields produced by the body.
- **Controlled Experiments**: Design double-blind, randomized controlled trials to test the effects of energy healing modalities on physiological and psychological outcomes. Control for placebo effects and ensure rigorous experimental protocols.
- **Interdisciplinary Collaboration**: Engage biophysicists, neuroscientists, and psychologists to develop standardized measures and methodologies for studying biofields.
- **Environmental Studies**: Examine the impact of environmental electromagnetic fields on human health and cognition, exploring how artificial and natural fields interact with biological systems (Belyaev, 2015).
- **Qualitative Studies**: Conduct phenomenological research capturing practitioners' and recipients' subjective

experiences of energy healing, providing insights into perceived effects and mechanisms.

Addressing Speculative Nature

Acknowledging the speculative aspects of Energetic Fields, it is essential to approach this layer with scientific rigor:

- **Critical Evaluation**: Recognize the limitations of current research and the need for replication and validation.
- **Open Inquiry**: Maintain openness to new evidence while applying strict methodological standards.
- **Transparency**: Clearly report experimental methods and results, allowing for peer review and scrutiny.
- **Ethical Considerations**: Ensure ethical practices in research involving human participants, including informed consent and safety measures.

Practical Applications

Applications of the Energetic Fields layer include:

- **Integrative Health Practices**: Incorporate energy-based therapies, such as Reiki, Healing Touch, and Qigong, into complementary medicine.
- **Stress Reduction**: Utilize practices that harmonize biofields, potentially reducing stress and enhancing well-being.
- **Environmental Design**: Consider the impact of electromagnetic environments on human health, promoting spaces that support biofield integrity.
- **Technological Innovations**: Develop wearable devices that monitor biofield activity for health assessment and personalized interventions.

2.3 Conscious Awareness

Definition and Theoretical Grounding

The **Conscious Awareness** layer encapsulates the subjective, reflective aspect of human experience. *Anthrocosmia* aligns with philosophical perspectives such as panpsychism, which posits that consciousness is a fundamental feature of the universe (Goff, Seager, & Allen-Hermanson, 2020), and Integrated Information Theory (IIT), which suggests that consciousness corresponds to the ability of a system to integrate information (Tononi, 2008).

In this view, humans are seen as focal points of cosmic awareness, allowing them to experience self-reflection, creativity, and interconnectedness with a broader consciousness beyond individual identity. This perspective challenges materialist views that reduce consciousness to neural processes, instead proposing that consciousness is an intrinsic aspect of reality.

The Conscious Awareness layer also considers the role of phenomenology—the study of structures of consciousness from a first-person perspective (Husserl, 2012). By acknowledging subjective experience as a valid source of knowledge, *Anthrocosmia* embraces a more holistic understanding of consciousness.

Integration with Existing Literature

Research on altered states of consciousness, such as those induced through meditation, psychedelics, or near-death experiences, provides insights into the nature of Conscious Awareness. Studies have shown that psychedelic substances like psilocybin can decrease activity in the brain's default mode network, leading to experiences of ego dissolution and interconnectedness (Carhart-Harris et al., 2012; Lebedev et al., 2015).

Mindfulness meditation has been associated with increased connectivity between brain regions involved in attention, self-regulation, and introspection (Tang et al., 2015). These findings suggest that consciousness can transcend ordinary states, accessing broader dimensions of awareness.

Transpersonal psychology explores experiences that transcend the usual limits of ego and individuality, including spiritual experiences, peak experiences, and mystical states (Friedman & Hartelius, 2015). This field provides a theoretical foundation for understanding Conscious Awareness as an expanded state of being.

Empirical Evidence and Methodological Proposals

Research on Conscious Awareness within *Anthrocosmia* could involve:

- **Neurophenomenology**: Integrate first-person subjective reports with third-person neural measurements to study consciousness (Varela, 1996).
- **Neuroimaging of Altered States**: Use fMRI, PET scans, and EEG to investigate brain activity during meditation, psychedelic experiences, and flow states.
- **Qualitative Research**: Employ phenomenological methods to capture the richness of subjective experiences, exploring themes of unity, transcendence, and interconnectedness.

- **Comparative Studies**: Examine cross-cultural accounts of consciousness to identify universal patterns and variations.
- **Psychometric Assessments**: Develop and utilize scales to measure aspects of consciousness, such as mindfulness, spiritual intelligence, and self-transcendence (King & DeCicco, 2009).

Practical Applications

Applications in this layer include:

- **Transpersonal Psychology**: Incorporate approaches that address spiritual and existential dimensions of human experience (Friedman & Hartelius, 2015).
- **Mindfulness-Based Interventions**: Use mindfulness and contemplative practices to enhance self-awareness, emotional regulation, and well-being.

- **Psychedelic-Assisted Therapy**: Explore therapeutic uses of psychedelics in controlled settings to address mental health conditions (Johnson & Griffiths, 2017).
- **Creativity Enhancement**: Apply techniques that expand consciousness to foster creativity and innovation in various fields.
- **Leadership Development**: Utilize practices that cultivate conscious awareness to develop mindful and ethical leadership.

Purpose and Potential of Human Existence

3.1 Human Beings as Learning Nodes in a Conscious Universe

Anthrocosmia posits that humans serve as "learning nodes" within a conscious universe, contributing to the cosmos's self-awareness through their experiences, emotions, and intentions. This idea aligns with concepts from process philosophy, which views reality as dynamic and evolving (Whitehead, 1978), and with theories of collective consciousness (Jung, 1969).

In this framework, human aspirations for knowledge, meaning, and connection are seen as expressions of the universe's intrinsic drive toward self-understanding. The interconnectedness of individuals suggests that personal growth and development contribute to a larger evolutionary process (Wilber, 2000).

3.2 Self-Development and Psychological Well-Being

By integrating the three layers, *Anthrocosmia* offers a holistic approach to self-development and psychological well-being:

- **Resilience and Adaptability**: Practices that align Biological Systems, Energetic Fields, and Conscious Awareness can enhance resilience and adaptability.
- **Emotional Regulation**: Mind-body interventions can improve emotional regulation by engaging physiological

processes and conscious reflection.
- **Meaning and Purpose**: Exploring existential questions within the framework can foster a sense of meaning and purpose.
- **Self-Transcendence**: Encouraging experiences that go beyond personal ego can lead to greater empathy, compassion, and social connectedness.

Empirical studies have shown that interventions promoting self-awareness and mindfulness can reduce symptoms of anxiety and depression, improve emotional regulation, and enhance overall well-being (Hofmann et al., 2010).

3.3 Collective Consciousness and Social Connectedness

The framework extends to collective dimensions:

- **Social Neuroscience**: Research shows that social interactions influence brain function and well-being

(Cacioppo & Cacioppo, 2012). Mirror neurons, for example, facilitate empathy and understanding of others' intentions (Rizzolatti & Sinigaglia, 2010).
- **Group Cohesion**: Practices like synchronized movement and group meditation can enhance social bonding and empathy (Tarr et al., 2016). Collective rituals and ceremonies have been found to strengthen group identity and cooperation (Xygalatas et al., 2013).
- **Global Consciousness**: The Global Consciousness Project suggests that collective human consciousness may influence physical systems (Nelson & Radin, 2003). While controversial, these findings invite further exploration of interconnectedness on a global scale.
- **Cultural Evolution**: The transmission of knowledge, beliefs, and practices contributes to the evolution of cultures and societies, reflecting the collective aspect

of human consciousness (Henrich, 2015).

3.4 Ethical Implications

Recognizing humans as interconnected participants in a conscious universe carries ethical implications:

- **Environmental Stewardship**: Acknowledging interconnectedness with the cosmos encourages responsible stewardship of the environment and sustainable practices.
- **Social Responsibility**: Understanding the impact of individual actions on the collective promotes ethical behavior and social justice.
- **Compassion and Empathy**: Cultivating conscious awareness can lead to increased compassion and efforts to alleviate suffering.

Applications and Implications of Anthrocosmia

4.1 Medicine

Integrative Health Approaches

- **Mind-Body Medicine**: Incorporate practices like MBSR, acupuncture, and biofeedback into treatment plans.
- **Energy Therapies**: Explore the efficacy of energy healing modalities in managing pain, anxiety, and other conditions.
- **Holistic Assessment**: Use comprehensive evaluations that consider physical, emotional, energetic, and conscious dimensions.

Empirical Evidence

- **Clinical Trials**: Studies show that integrative therapies can reduce symptoms and improve quality of life (Sherman et al., 2013).
- **Physiological Markers**: Research indicates that practices like meditation can modulate stress responses and immune function (Black & Slavich, 2016).
- **Cost-Effectiveness**: Integrative approaches may

reduce healthcare costs by preventing illness and reducing the need for invasive treatments (Herman et al., 2012).

Challenges and Considerations

- **Standardization**: Developing standardized protocols for integrative therapies is necessary for widespread adoption.
- **Training**: Healthcare professionals require training in holistic practices and cultural competencies.
- **Patient Acceptance**: Educating patients about the benefits and evidence supporting integrative approaches can enhance acceptance.

4.2 Psychology

Holistic Therapeutic Practices

- **Integrative Psychotherapy**: Combine cognitive-behavioral techniques with somatic and mindfulness approaches.

- **Energy Psychology**: Techniques like Emotional Freedom Techniques (EFT) have shown efficacy in reducing anxiety and PTSD symptoms (Church et al., 2017).
- **Transpersonal Therapy**: Address spiritual and existential dimensions, facilitating self-transcendence and meaning-making.

Empirical Evidence

- **Meta-Analyses**: Reviews demonstrate the effectiveness of mindfulness-based interventions for various mental health conditions (Goldberg et al., 2018).
- **Neural Correlates**: Studies link therapeutic practices to changes in brain structure and function (Holzel et al., 2011).
- **Cultural Sensitivity**: Incorporating clients' cultural and spiritual beliefs enhances therapeutic outcomes (Sue et al., 2009).

Challenges and Considerations

- **Integration into Mainstream Practice**: Overcoming skepticism and resistance within the professional community.
- **Ethical Practice**: Ensuring that interventions are evidence-based and ethically applied.
- **Accessibility**: Making holistic therapies accessible to diverse populations.

4.3 Artificial Intelligence (AI)

Ethically Conscious AI Development

- **Human-Centered Design**: Develop AI systems that enhance well-being, empathy, and social connection.
- **AI in Mental Health**: Use AI-driven platforms to provide accessible mental health support while ensuring ethical considerations.
- **Augmented Intelligence**: AI can augment human capabilities rather than replace them, supporting decision-making and creativity.

Empirical Evidence

- **AI Therapists**: Early studies show that AI chatbots can reduce symptoms of depression and anxiety (Fitzpatrick et al., 2017).
- **Ethical Frameworks**: Research emphasizes the importance of ethics in AI development to prevent biases and promote positive outcomes (Floridi et al., 2018).
- **Social Robotics**: Robots designed for social interaction can support education and therapy (Wada & Shibata, 2007).

Challenges and Considerations

- **Ethical AI**: Ensuring that AI systems are transparent, fair, and respect privacy.
- **Human-AI Interaction**: Understanding the psychological impact of interacting with AI.
- **Technological Access**: Addressing the digital divide to ensure equitable access to AI technologies.

4.4 Education

Holistic Educational Models

- **Social-Emotional Learning (SEL)**: Incorporate SEL programs to develop empathy, self-awareness, and relationship skills.
- **Contemplative Education**: Use mindfulness and reflective practices to enhance focus and well-being.
- **Integrative Curriculum**: Combine academic learning with experiential, artistic, and ecological education.

Empirical Evidence

- **Academic Performance**: Studies link SEL to improved academic outcomes and reduced behavioral problems (Durlak et al., 2011).
- **Cognitive Benefits**: Mindfulness training in schools has been associated with better attention and emotional regulation (Zenner et al., 2014).

- **Long-Term Outcomes**: Holistic education may contribute to lifelong well-being and social responsibility (Steiner, 1995).

Challenges and Considerations

- **Curriculum Development**: Designing programs that meet educational standards while incorporating holistic principles.
- **Teacher Training**: Educators require training in holistic practices and methods.
- **Assessment**: Developing metrics to evaluate holistic education's impact.

Addressing Potential Criticisms and Limitations

5.1 Speculative Nature of Energetic Fields

Criticism: The concept of Energetic Fields lacks robust empirical support and may be considered pseudoscientific.

Response:

- **Emerging Research**: Acknowledge that biofield science is an emerging field with preliminary evidence. Emphasize the need for rigorous, controlled studies.
- **Interdisciplinary Collaboration**: Advocate for collaborative research across disciplines to develop standardized methodologies.
- **Open Inquiry**: Maintain openness to refutation, aligning with scientific principles.
- **Historical Precedents**: Note that concepts like neuroplasticity were once controversial but gained acceptance through empirical validation.

5.2 Panpsychism and Consciousness

Criticism: Panpsychism is a controversial view that may not be empirically testable.

Response:

- **Theoretical Value**: Highlight that panpsychism offers a

valuable philosophical perspective that addresses the hard problem of consciousness (Chalmers, 1996).
- **Integration with Science**: Point to efforts to develop testable predictions based on panpsychism (Goff et al., 2020).
- **Alternative Theories**: Acknowledge alternative views and encourage comparative studies.
- **Phenomenological Evidence**: Emphasize the importance of subjective experience as valid data.

5.3 Methodological Challenges

Criticism: Studying subjective experiences and subtle phenomena poses methodological challenges.

Response:

- **Mixed Methods**: Propose using both quantitative and qualitative methods to capture the richness of human experience.

- **Advancements in Technology**: Utilize emerging technologies and innovative research designs.
- **Replication and Transparency**: Encourage replication studies and transparent reporting to build a robust evidence base.

5.4 Cultural and Philosophical Biases

Criticism: The framework may be influenced by cultural or philosophical biases that limit its universal applicability.

Response:

- **Cultural Inclusivity**: Incorporate cross-cultural perspectives and recognize the diversity of human experiences.
- **Interdisciplinary Dialogue**: Engage scholars from various traditions to enrich the framework.
- **Universal Themes**: Identify universal aspects of human

experience while respecting cultural differences.

Conclusion and Future Directions

6.1 Contributions and Impact

Anthrocosmia offers a novel, integrative framework that redefines human existence within a cosmic context. By incorporating biological, energetic, and conscious dimensions, it bridges scientific inquiry with philosophical depth, addressing both empirical and existential questions.

This framework contributes to:

- **Theoretical Innovation**: Proposes a comprehensive model that integrates multiple dimensions of human experience.
- **Interdisciplinary Integration**: Bridges gaps between neuroscience, psychology, biophysics, and philosophy.
- **Practical Applications**: Provides actionable insights for medicine, psychology, AI, and education.

6.2 Future Research Avenues

- **Empirical Validation**: Conduct rigorous studies to test the framework's concepts, particularly in Energetic Fields and Conscious Awareness.
- **Interdisciplinary Collaboration**: Encourage partnerships among neuroscientists, psychologists, biophysicists, philosophers, and practitioners.
- **Application Development**: Develop interventions, educational programs, and technologies based on the framework.
- **Policy Implications**: Explore how the framework can inform policy in health care, education, and technology ethics.

6.3 Final Remarks

By honoring the complexity of human consciousness and existence, *Anthrocosmia* invites a paradigm shift toward a more holistic, interconnected understanding of humanity's role in a conscious universe. As empirical evidence accumulates, this framework has the potential to contribute significantly to psychology, consciousness studies, and beyond.

Embracing *Anthrocosmia* can lead to:

- **Enhanced Well-Being**: Holistic practices that improve individual and collective health.
- **Ethical Development**: Technologies and policies that reflect human values and interconnectedness.
- **Global Harmony**: Greater understanding and cooperation across cultures and societies.

Works Cited

Foundational Scientific Theories and Quantum Mechanics

1. Aspect, A., Dalibard, J., & Roger, G. (1982). *Experimental Test of Bell's Inequalities Using Time-Varying Analyzers*. *Physical*

Review Letters, 49(25), 1804-1807. doi:10.1103/PhysRevLet t.49.1804.
2. Bell, J. S. (1964). *On the Einstein Podolsky Rosen Paradox. Physics Physique Физика*, 1(3), 195-200.
3. Bohm, D. (1980). *Wholeness and the Implicate Order*. London: Routledge.
4. Einstein, A. (1954). *Ideas and Opinions*. New York: Crown Publishers.
5. Penrose, R., & Hameroff, S. R. (1996). *Consciousness in the Universe: Neuroscience, Quantum Space-Time Geometry and Orch-OR Theory. Journal of Cosmology*, 14, 1-17.
6. Schrödinger, E. (1935). *Discussion of*

Probability Relations between Separated Systems. Proceedings of the Cambridge Philosophical Society, 31, 555-563.

Neuroscience and Microtubules

1. Hameroff, S., & Penrose, R. (1996). *Orchestrated Reduction of Quantum Coherence in Brain Microtubules: A Model for Consciousness. Mathematics and Computers in Simulation*, 40(3), 453-480.
2. Hasson, U., & Malach, R. (2004). *Brain-to-Brain Synchronization during Natural Communication. Proceedings of the National Academy of*

Sciences, 101(23), 7635-7639.
3. HeartMath Institute. (2017). *The Science of Heart Intelligence: Exploring the Role of the Heart in Human Performance*. Boulder Creek: HeartMath Institute.
4. Jibu, M., & Yasue, K. (1995). *Quantum Brain Dynamics and Consciousness: An Introduction*. Amsterdam: John Benjamins Publishing.

Collective Consciousness and Social Theories

1. Jung, C. G. (1960). *The Collected Works of C.G. Jung: Volume 9 (Part 1): The Archetypes and the Collective Unconscious*. Princeton,

NJ: Princeton University Press.
2. Jung, C. G., & Pauli, W. (1955). *The Interpretation of Nature and the Psyche*. New York: Pantheon.
3. Sheldrake, R. (1981). *A New Science of Life: The Hypothesis of Morphic Resonance*. Los Angeles: Tarcher.
4. Yates, J. F., & McLean, M. R. (2006). *The Global Consciousness Project: A Study of Collective Emotional Reactions to the September 11, 2001, Terrorist Attacks. Journal of Parapsychology*, 70(1), 121-137.

Resonance, Frequency, and Cymatics

1. Emoto, M. (2004). *The Hidden Messages in Water*. New York: Beyond Words Publishing.
2. Jenny, H. (2001). *Cymatics: A Study of Wave Phenomena & Vibration*. Newmarket, NH: MACROmedia Publishing.
3. Tesla, N. (1893). *The Transmission of Electric Energy Without Wires. Electrical World and Engineer*, 5, 429-431.

The Observer Effect and Mind-Matter Interaction

1. Bohr, N. (1934). *Atomic Theory and the Description of Nature*. Cambridge: Cambridge University Press.
2. Radin, D. (1997). *The Conscious Universe:*

The Scientific Truth of Psychic Phenomena. New York: HarperEdge.
3. Targ, R., & Puthoff, H. E. (1974). *Information Transmission under Conditions of Sensory Shielding*. *Nature*, 252(5481), 602-607.

Non-Locality and Quantum Mind Theories

1. Libet, B. (1985). *Unconscious Cerebral Initiative and the Role of Conscious Will in Voluntary Action. Behavioral and Brain Sciences*, 8(4), 529-566.
2. Laszlo, E. (2007). *Science and the Akashic Field: An Integral Theory of Everything*. Rochester, VT: Inner Traditions.

Consciousness Studies and Philosophy

1. Chalmers, D. J. (1995). *Facing up to the Problem of Consciousness. Journal of Consciousness Studies*, 2(3), 200-219.
2. James, W. (1890). *The Principles of Psychology*. New York: Henry Holt.
3. Planck, M. (1944). *Das Wesen der Materie* [The Nature of Matter]. Lecture to the German Physical Society.

Empirical Studies and Case Studies on Consciousness Effects

1. Radin, D., & Nelson, R. (2002). *Exploring Relationships between Random Physical Events and Mass Human*

Attention: Asking for Whom the Bell Tolls. Journal of Scientific Exploration, 16(4), 579-596.

2. Travis, F., & Shear, J. (2010). *Focused Attention, Open Monitoring, and Automatic Self-Transcending: Categories to Organize Meditations from Vedic, Buddhist, and Chinese Traditions. Consciousness and Cognition*, 19(4), 1110-1118.

Sound and Light Therapies

1. Goldstein, A. V., & Ciorba, C. (2018). *The Effectiveness of Sound Therapy on Tinnitus: A Review of Literature. Journal of Clinical*

Audiology, 11(3), 187-192.
2. Schwenger, C., & Ross, J. M. (2019). *Light Therapy for Seasonal Affective Disorder: A Review of Treatment Options. Journal of Clinical Psychology*, 75(2), 146-159.

Biofeedback and Heart Coherence

1. Lehrer, P. M., & Gevirtz, R. (2014). *Heart Rate Variability Biofeedback: How and Why Does It Work?. Frontiers in Psychology*, 5, 756. doi:10.3389/fpsyg.2014.00756.
2. McCraty, R., Atkinson, M., & Tomasino, D. (2001). *Science of the Heart: Exploring the Role of the Heart in*

Human Performance. Boulder Creek, CA: HeartMath Research Center, Institute of HeartMath.

Group Intention Studies and Random Event Generators

1. Nelson, R. D., & Radin, D. I. (2004). *Field REG II: Consciousness Field Effects: Replications and Explorations. Journal of Scientific Exploration*, 18(2), 219-241.
2. Jahn, R. G., & Dunne, B. J. (1987). *Margins of Reality: The Role of Consciousness in the Physical World.* New York: Harcourt Brace Jovanovich.

Educational Theory and Applications of Collective Consciousness

1. Dewey, J. (1938). *Experience and Education*. New York: Collier Books.
2. Vygotsky, L. S. (1978). *Mind in Society: The Development of Higher Psychological Processes*. Cambridge, MA: Harvard University Press.

Technology and AI in Relation to Consciousness

1. Kurzweil, R. (2005). *The Singularity is Near: When Humans Transcend Biology*. New York: Viking Penguin.
2. Tegmark, M. (2017). *Life 3.0: Being Human in the Age of Artificial Intelligence*. New York: Knopf.

Environmental Impact and Mindful Conservation

1. Austin, V. (2021). *The Secret Intelligence of Water: Macroscopic Evidence of Water Responding to Human Consciousness.* New York: Atria Books.
2. Lovelock, J. (2000). *Gaia: A New Look at Life on Earth.* Oxford: Oxford University Press.

Additional References for Future Directions and Applications

1. Loye, D. (2004). *The Great Adventure: Toward a Fully Human Theory of Evolution.* Albany, NY: SUNY Press.
2. Whitehead, A. N. (1929). *Process and Reality.* New York: Macmillan.

CIA Documents and Government Research Programs

1. Central Intelligence Agency. *Analysis and Assessment of Gateway Process.* Declassified Document, date unknown. Retrieved from https://www.cia.gov/readingroom/document/0000017788
2. Central Intelligence Agency. *Project MK-Ultra, the CIA's Program of Research in Behavioral Modification.* Declassified Document, date unknown. Retrieved from https://www.cia.gov/readingroom/document/0000145476
3. Central Intelligence Agency. *The Stargate Project: An Evaluation of Remote Viewing Research and*

Applications. Declassified Document, date unknown. Retrieved from https://www.cia.gov/readingroom/document/0000083943

Color Theory References

1. Chevreul, M.-E. (1839). *The Principles of Harmony and Contrast of Colors.* London: Longman, Brown, Green, and Longmans.
2. Eliasson, O. (2019). *Experience.* New York: Phaidon Press.
3. Gayford, M., & Turner, J. (2018). *Olafur Eliasson: In Real Life.* London: Tate Publishing.
4. Goethe, J. W. (1970 [1810]). *Theory of Colours.* Trans. C.

Eastlake. Cambridge, MA: MIT Press.
5. Hering, E. (1964 [1878]). *Outlines of a Theory of the Light Sense.* Cambridge, MA: Harvard University Press.
6. Kawabata, H., & Zeki, S. (2004). *Neural Correlates of Beauty. Journal of Neurophysiology, 91*(4), 1699–1705. https://doi.org/10.1152/jn.00696.2003.
7. Munsell, A. H. (1905). *A Color Notation.* Boston: Geo. H. Ellis Co.
8. Newton, I. (1979 [1704]). *Opticks.* New York: Dover Publications.
9. Turrell, J. (2009). *The Art of Light and Space.*

Berkeley: University of California Press.
10. Ward, J. (2004). *The Frog Who Croaked Blue: Synesthesia and the Mixing of the Senses.* London: Routledge.
11. Ward, J., & Meijer, P. (2010). *Visual Experiences in the Blind Induced by an Auditory Sensory Substitution Device. Consciousness and Cognition, 19*(1), 492–500. https://doi.org/10.1016/j.concog.2010.01.006.
12. Wilber, K. (2001). *A Theory of Everything: An Integral Vision for Business, Politics, Science, and Spirituality.* Boston: Shambhala Publications.
13. Zeki, S. (1999). *Inner Vision: An Exploration of Art and*

the Brain. Oxford: Oxford University Press.

Anthrocosmia References

1. Barsalou, L. W. (2008). *Grounded cognition. Annual Review of Psychology, 59,* 617–645. https://doi.org/10.1146/annurev.psych.59.103006.093639
2. Becker, R. O., & Selden, G. (1985). *The Body Electric: Electromagnetism and the Foundation of Life.* William Morrow.
3. Belyaev, I. (2015). *Biophysical mechanisms for nonthermal microwave effects. Bioelectromagnetics, 36*(5), 393–413. https://doi.org/10.1002/bem.21920

4. Black, D. S., & Slavich, G. M. (2016). *Mindfulness meditation and the immune system: a systematic review of randomized controlled trials. Annals of the New York Academy of Sciences, 1373*(1), 13–24. https://doi.org/10.1111/nyas.12998
5. Cacioppo, S., & Cacioppo, J. T. (2012). *Decoding the invisible forces of social connections. Frontiers in Integrative Neuroscience, 6,* 51. https://doi.org/10.3389/fnint.2012.00051
6. Carhart-Harris, R. L., et al. (2012). *Neural correlates of the psychedelic state as determined by fMRI studies with psilocybin.*

Proceedings of the National Academy of Sciences, 109(6), 2138–2143. https://doi.org/10.1073/pnas.1119598109
7. Chalmers, D. J. (1995). *Facing up to the problem of consciousness. Journal of Consciousness Studies, 2*(3), 200–219.
8. Chalmers, D. J. (1996). *The Conscious Mind: In Search of a Fundamental Theory.* Oxford University Press.
9. Church, D., & Feinstein, D. (2017). *Energy psychology: Efficacy, speed, mechanisms. Explore, 13*(6), 411–418. https://doi.org/10.1016/j.explore.2017.08.003
10. Crick, F., & Koch, C. (2003). *A framework

for consciousness. Nature Neuroscience, 6(2), 119–126. https://doi.org/10.1038/nn0203-119

11. Davidson, R. J., & McEwen, B. S. (2012). *Social influences on neuroplasticity: stress and interventions to promote well-being. Nature Neuroscience, 15*(5), 689–695. https://doi.org/10.1038/nn.3093

12. Dehaene, S. (2014). *Consciousness and the Brain: Deciphering How the Brain Codes Our Thoughts*. Viking.

13. Durlak, J. A., et al. (2011). *The impact of enhancing students' social and emotional learning: A meta-analysis. Child*

Development, 82(1), 405–432. https://doi.org/10.1111/j.1467-8624.2010.01564.x
14. Edelman, G. M. (2004). *Wider than the Sky: The Phenomenal Gift of Consciousness.* Yale University Press.
15. Ferrer, J. N. (2002). *Revisioning Transpersonal Theory: A Participatory Vision of Human Spirituality.* SUNY Press.
16. Friedman, H. L., & Hartelius, G. (Eds.). (2015). *The Wiley-Blackwell Handbook of Transpersonal Psychology.* Wiley-Blackwell.
17. Goff, P., Seager, W., & Allen-Hermanson, S. (2020). *Panpsychism.* In E. N.

Zalta (Ed.), *The Stanford Encyclopedia of Philosophy* (Summer 2020 ed.). https://plato.stanford.edu/entries/panpsychism/

1. Goldberg, S. B., et al. (2018). *Mindfulness-based interventions for psychiatric disorders: A systematic review and meta-analysis.* Clinical Psychology Review, 59, 52–60. https://doi.org/10.1016/j.cpr.2017.10.011
2. Hall, H., et al. (2016). *Energy medicine: Rejection of a homeopathic energy approach in cancer treatment.* Current Oncology, 23(3), e179–e181. https://doi.org/10.3747/co.23.3138

3. Henrich, J. (2015). *The Secret of Our Success: How Culture Is Driving Human Evolution, Domesticating Our Species, and Making Us Smarter.* Princeton University Press.
4. Herman, P. M., et al. (2012). *Are complementary therapies and integrative care cost-effective? A systematic review of economic evaluations. BMJ Open, 2*(5), e001046. https://doi.org/10.1136/bmjopen-2012-001046
5. Hofmann, S. G., et al. (2010). *The effect of mindfulness-based therapy on anxiety and depression: A meta-analytic review. Journal of Consulting and Clinical Psychology,*

78(2), 169–183. https://doi.org/10.1037/a0018555

6. Holzel, B. K., et al. (2011). *Mindfulness practice leads to increases in regional brain gray matter density.* Psychiatry Research: Neuroimaging, *191*(1), 36–43. https://doi.org/10.1016/j.pscychresns.2010.08.006

www.ingramcontent.com/pod-product-compliance
Lightning Source LLC
Chambersburg PA
CBHW031610210526
45464CB00004B/1517